Berries

and other small fruit

Berries

and other small fruit

All the pastry recipes were created by Patrice Demers
Photographer: Pierre Beauchemin

Jean-Paul Grappe

Fitzhenry & Whiteside

Contents

First published in English by Fitzhenry & Whiteside in 2014

www.fitzhenry.ca godwit@fitzhenry.ca

Fitzhenry & Whiteside acknowledges with thanks the Canada Council for the Arts, and the Ontario Arts Council for their support of our publishing program. We acknowledge the financial support of the Government of Canada through the Canada Book Fund (CBF) for our publishing activities.

Canada Council for the Arts Conseil des Arts du Canada

Translator: Marcella Walton
Designer: Daniel Choi

Library and Archives Canada Cataloguing in Publication
Grappe, Jean-Paul
[Petits fruits. English]
Berries and Other Small Fruit / Jean-Paul Grappe.
Translation of: Petits fruits.
ISBN 978-1-55455-287-0 (pbk.)
1. Cooking (Berries). 2. Berries. 3. Cookbooks. I. Title.
II. Title: Petits fruits. English.
TX813.B4G7213 2014 641.6'47 C2013-908195-X

Publisher Cataloging-in-Publication Data (U.S.)
Berries and Other Small Fruit
ISBN 978-1-55455-287-0
Data available on file

Printed and bound by Sheck Wah Tong Printing Press Ltd. in Hong Kong, China
10 9 8 7 6 5 4 3 2 1

The team, from left to right: Patrice Demers, pastry chef and co-owner of restaurants *Les Chèvres* and *Le Chou*;
Pierre Beauchemin, food photographer, ITHQ;
Jean-Paul Grappe, chef de cuisine and teacher, ITHQ;
Julien Bartoluci, illustrator;
Colombe St. Pierre, food stylist, chef de cuisine and owner of *Chez St. Pierre*, au Bic;
Lucie St. Gelais, assistant stylist.

Introduction

When the first explorers discovered Quebec, they were surprised to find many varieties of vegetation and as many interesting ways of eating them. The Native peoples found all the nutritional and medicinal ingredients they needed in the plants around them. The fact is that the province of Quebec is so large that wild plants and fruit vary greatly from one region to another. For example, there are no cloudberries in Montreal nor are there black raspberries in the Gaspé. The well-known botanist Frère Marie-Victorin first identified Quebec's wild plants and fruit in his *Flore laurentienne*, published in 1935.

Today, Quebecois cuisine has enthusiastically rediscovered all of these indigenous plants and fruits, while speciality growers entice us into wonderful culinary adventures.

In July, we can't wait to taste the strawberries, but did you know that there are even more delectable small, wild strawberries growing on our hills and in our valleys everywhere in Quebec? Let's take advantage! In the fall, we can harvest lingonberries and both small and large cranberries, but these berries become even more juicy and tasty after a few frosts. Cloudberries, which make Chicoutai liqueur, are a kind of white raspberry that is found in the higher latitudes of Nordic countries. They are delicious in sorbets or cakes. High bush cranberries, dwarf raspberries, wild cherries, ground cherries, rowanberries, and flowering raspberries are among those berries that are a delight for foragers.

Today, fruit comes to us from all over the world thanks to modern transportation. We don't even have to wait for strawberry season in Quebec. We can find them on our grocery shelves all year round. Sometimes they are good and sometimes they're not so good. Although lychees do not grow here, they are highly prized—juicy, sweet, delicious, and a delight for the taste buds. It's important to me to cook with the fruit found in our local markets, but also to understand its origins.

Fruit-tree arboriculture was born through the propagation of cutting, layering, and grafting, through the selection of the best varieties of a species, and the creation of cultivars (artificially obtained varieties). At one time, the Sumerians of Mesopotamia farmed the date palm, which may have originated near the Dead Sea. It became the first cultivated tree about six or seven thousand years ago. According to the Bible, the Hebrews cultivated olive, almond, pistachio, carob bean trees, and vineyards in Judea. Peach and apricot trees were grown for centuries in China.

When I was young, we ate fruit when it was very ripe. It was bruised in places because it was fragile, but it was incredibly delectable. These days, because of the distances it must travel, we often eat tasteless fruit that isn't ripe. Who prefers bananas when the skin is black or spotted? Almost no one. But they taste best at this stage. Unfortunately, everything is affected by this. Yellow peaches, nectarines, and red plums, are created in laboratories so they almost never ripen properly, which means they slowly spoil instead.

In Quebec, we have "le merisier," a wild red cherry. My son picks these wild cherries for me to cook with every year because they're better—not as sweet as cherries found in grocery stores. Wild cherries are beautifully suited to duck, for example, or else you can substitute sour Morello or Montmorency cherries. They are essential to certain classical recipes; if they're not available, you have to make a gastrique, (a caramel prepared with sugar and an acidic ingredient that can be incompatible with certain meats or poultry).

I have been cooking professionally for forty-eight years and, for a long time, I was against cooking with fruit. However, staying with Native peoples and, above all, using wild fruit, made me realize that fruit used to be far less sweet. That's why the idea of writing this book ignited my passion. In careful combinations, we can boost the flavour of basic ingredients by the addition of certain fruits and vegetables, in what could be called a happy marriage.

I do not cook game meat unless I know the animal's diet. For example, the willow ptarmigan feeds on small green shoots but the rock ptarmigan feeds on berries found in the tundra, like crowberries and small cranberries. So I cook rock ptarmigan with little cranberries and insist on serving it with a wine that has a fruity bouquet!

Recently, during a trip to Île aux Coudres, I talked to an apple grower, Mr. Pedneault, about heirloom apples from the past. We used to grow so many varieties: the Snow Apple ("Fameuse"), very sweet; the Wealthy, slightly acidic; the Alexander, a cooking apple; the "betpel"; the Greening, that doesn't ripen on the tree; the Duchess; etc. On our store shelves, however, the apples we find are certainly good, but "factory produced." The Golden Delicious, Cortland, Reinette Russet, and Granny Smith apples keep for a long time, and the Geneva, inedible when raw, yields a very nice rose cider. Plums once were also abundant everywhere in Quebec. Damas plums were saved by Paul-Louis Martin. When he and his wife bought the former estate of Sifroy Guéret in 1975, they restored an abandoned orchard and founded the Maison de la Prune in Saint André in Kamouraska.

Cooking with plums is an adventure, and mirabelles, greengages, and damsons allow us to discover some interesting dishes. I also like to use fruit that is less familiar here, for example, quince, olives, and figs, as well as strawberries—even though they are not fruits. Surprising, isn't it? The actual fruit of the strawberry plant is the achene, the hard little seeds on the outside of the strawberry, that are crunchy when chewed. As for the red pulp, it is the floral receptacle. Figs are even more surprising—they are not a fruit but a flower that becomes a fleshy receptacle that encloses the fruit. Inside the fig, the female flower bears a small, dry fruit that is used for seeds. While strawberries and figs are not fruit, tomatoes, eggplants, squash, cucumber, peppers, and chili peppers are. Laymen and botanists are not always in agreement, but who is right?

Whatever the answer, it is only through cooking that we know the joy of eating.

Degree of difficulty of the recipes:
1 Easy
2 Somewhat easy
3 Medium
4 Somewhat difficult
5 Difficult

Berries and Other Small Fruit

Apricots

Prunus armeniaca (Apricot tree)
Family: *Rosaceae*

Apricot trees are small with reddish bark and are six to eight metres high. Their attractive, fragrant flowers are a corolla of white, tinted with pink. They appear before the leaves, but can't withstand spring frosts. Apricots have been grown since antiquity. Research shows that the Chinese have grown them for 4000 years. From the Far East, the apricot travelled through western Asia to Armenia, hence the name of the species: *armeniaca*. Today, it is one of the most cultivated fruit trees, for eating as well as for preserving. The kernel of the pit, which is very oily, is edible when soft, but more often than not it is bitter.

But this fuzzy yellow-orange fruit that is now considered such a treat, struggled to find favour. Historically, raw apricot was accused of having harmful effects. Indeed, until the 18th century, it was recommended only for compotes or jams. On the other hand, the oil extracted from the kernel has a number of extraordinary benefits. When underripe, the flesh of the apricot is rubbery; overripe and it can leave a lingering feel in the mouth of squashed bugs.

Cooking and Baking

Apricots can enhance poultry and salads. Use in compotes, jams, cakes, jellies, mousses, fruit paste, sorbets, pies.

Therapeutic Uses

Anemia, aperitif, astringent, immunity, laxative, nutrient, cooling agent, tissue regenerator.

Rabbit Thighs and Apricots

4 servings • Cooking time: 40 minutes • 3

Preparation: 20 minutes

- Maceration: place the apricots, wine, liqueur, and lemon juice in a bowl. Cover with a lid and macerate for at least 8 days.
- Get your butcher to remove the bones from the rabbit thighs without cutting them open. Make sure to keep the bones and ask your butcher for more to make the rabbit stock.*
- Season the thighs with salt and pepper. Set aside. Drain the macerated apricots and finely mince them. Add them to the rabbit stuffing with the maceration juice. Generously stuff the thighs and wrap each one in caul fat.
- In a sauté pan, heat the oil and butter and sear the thighs on each side. Then put them in the oven at 400 °F (200 °C), basting occasionally. Halfway through cooking, when the thighs have reached 140 °F (60 °C), add the mirepoix. Cook again until the thighs reach an internal temperature of 175 °F (80 °C). At this point, remove the thighs and keep them warm. Pour off the excess cooking fat. Add the rabbit stock and cook for 5 to 8 minutes and strain. Adjust the seasoning and finish the jus with butter.
- Lay the thighs onto a bed of pasta and sauce with the rabbit jus.

* Use the same method as for brown poultry stock (see p. 130).

INGREDIENTS

- 4 rabbit thighs
- Salt and freshly ground pepper
- 7 oz (200 g) rabbit stuffing (recipe on page 125)
- 2 pieces of caul fat
- 1/4 cup (60 ml) cooking oil
- 1/4 cup (60 g) unsalted butter
- 5 oz (150 g) mirepoix (a mixture of diced onion, carrot, and celery)
- 1 cup (250 ml) unthickened brown rabbit stock or store-bought stock
- 1/3 cup (80 g) unsalted butter

Maceration

- 5 oz (150 g) dried apricots
- 6 tablespoons (90 ml) dry white wine
- 6 tablespoons (90 ml) apricot liqueur
- Juice of one lemon

Sautéed Poultry Gizzards Vol-au-Vents with Apricot and Cognac Sauce

4 servings • Cooking time: 40 minutes • 3

The taste of gizzards tends to be underrated. They have a very pronounced poultry flavour.

INGREDIENTS

- 16 dried apricots
- 1 cup (250 ml) dry white wine
- 6 tablespoons (90 ml) cognac
- 1 1/4 cups (160 g) gizzards, chicken, duck, or guinea fowl
- Salt and freshly ground pepper
- 7 oz (200 g) pearl onions
- 2 cloves of garlic
- 1 bouquet garni
- 1/3 cup (80 g) unsalted butter
- 1/3 cup (80 ml) cooking oil
- 4 carrots, cubed
- 2 cups (500 ml) thickened brown chicken stock
- 4 store-bought vol-au-vents or pastry cups

Preparation: 30 Minutes

- Macerate the dried apricots with the wine and cognac for a few hours.
- Wash the gizzards thoroughly, removing the rough parts. Season them with salt and pepper and set aside.
- Peel the small onions. Cut them in cubes and set aside. Finely chop the cloves of garlic and prepare the bouquet garni.
- In a heavy-bottom frying pan, heat the butter and cooking oil. Quickly sear the gizzards and put them in a saucepan. Sear the onions, garlic, and carrots and place them on top of the gizzards. Pour in the maceration liquid and simmer for a few minutes. Add the poultry stock and the bouquet garni. Cook slowly. Once the gizzards are tender, adjust the seasoning.
- Open the vol-au-vents, fill with the sautéed ingredients, and replace the top.

Apricot Ratafia Liqueur

- 8 oz (240 g) apricot seed kernels, crushed
- 4 cups (1 litre) white rum or eau-de-vie, 45%
- 2 egg whites
- 2 cups (500 g) granulated white sugar

Preparation: 20 Minutes

- Put the apricot kernels in a large-mouth Mason jar and cover with the white rum. Close the jar and macerate for 3 weeks. Decant and strain through a coffee filter. Let sit for 24 hours, decant and filter again.
- Make a syrup with the granulated sugar dissolved in a bit of boiling water. Cool. Add the syrup to the macerated ingredients and finish with enough water to make 4 cups (1 litre) of liqueur. Shake well. Let age for a few months. Ratafias made with cherry pits, peach, and plum stones are all prepared the same way.

Millefeuille with Apricots and Earl Grey Tea Custard

4 servings • Cooking time: 20 minutes • 3

Preparation: 30 Minutes

- Bring the milk and half of the sugar to a boil. Take it off the heat, add the tea leaves, and infuse for 5 minutes. Strain the milk through a cone-shaped mesh sieve. Cream the egg yolks with the rest of the sugar until light in colour and add the cornstarch. Bring the milk to a boil again and pour it over the yolks. Put it back in the saucepan and cook for 1 minute. Add the gelatin that has been "bloomed" (soaked) in water and drained. Pour the custard into a double boiler and gently heat, stirring regularly. Whip the cream to soft peaks and add this to the heated custard. Chill for a few hours.

Puff Pastry

- Roll out the store-bought puff pastry to 1/2" (1 cm) thickness. Chill the pastry on a cookie sheet in the fridge for at least 1 hour. Bake the pastry (with a weight on top to prevent it from rising too much) at 375 °F (190 °C) for about 20 minutes. Chill the pastry before cutting it.

Assembly for 1 serving

- Cut the pastry into 3 equal rectangles and the apricots into quarters. Using a pastry bag, pipe the custard on 2 of the pastry rectangles. Garnish with the apricots. Place one rectangle on the other and top with the third. Dust with icing sugar.

INGREDIENTS

Tea Custard

- 1 cup (250 ml) milk
- 1/4 cup (60 g) sugar
- 6 g (1 tsp) Earl Grey tea leaves
- 2 egg yolks
- 20 g (4 tsp) cornstarch
- 1 gelatin sheet
- 1/2 cup (125 ml) heavy cream (35%)
- 10 oz (300 g) puff pastry

Assembly

- Apricots
- Icing sugar

Blackberries

Rubus occidentalis
Family: *Rosaceae*

The common blackberry (*Rubus fruticosus*) is a shrub found in glades and the edge of woods. The elmleaf blackberry (*Rubus ulmifolius*) is found in very sunny areas. The berries are small and sweet. Even better are the blackberries from the dewberry shrub (*Rubus caesius*), a plant that likes cool and humid areas. Hybridization of raspberries and California blackberries (*Rubus vitifolius*) has created the loganberry. Crossing that with another plant from the genus *Rubus* makes the boysenberry, grown for its large and fragrant berries. Prehistoric man enjoyed blackberries in jams, compotes, and as a herbal astringent. Teas made from blackberry and raspberry leaves are delicious. Cloudberry (*Rubus chamaemorus*), sometimes called bake-apple, is part of this large family of blackberries.

Cooking and Baking

Alcohols or vinegars that have macerated blackberries can be used to deglaze dishes.

This fruit goes well with game meats and some fish.

Compotes, jams, juices, marmalades.

Cakes and pies.

Therapeutic Uses

Internal uses: pulmonary ailments, asthenia (weakness), constipation, hemorrhagic diathesis (bleeding tendency), enteritis.

External uses: angina, aphtha (canker), stomatitis (mouth inflammation).

Rose and Blackberry Floating Island

4 servings • Cooking time: 10 minutes

Preparation: 30 Minutes

- Boil the milk with the rose water.
- Beat the egg whites with the salt. Gradually incorporate the sugar. The whites should have stiff peaks. Place a spoonful of the whites into the boiling milk. Cook for about 2 minutes, turning it over halfway through cooking.
- Drain the cooked egg white and repeat these steps with the rest of the egg whites.
- Boil the rose-water milk with 1/3 cup (80 g) of sugar. Cream the egg yolks with the rest of the sugar. Gradually pour the boiling milk over the yolks while stirring.
- Put this mixture back into the saucepan on medium heat and cook until it coats the back of a spoon. Chill it over ice. Pour the custard into bowls or glasses.
- Add the blackberries and finish with the egg-white floating islands.

INGREDIENTS

Floating Island

- 3 cups (750 ml) milk
- 1 teaspoon (5 ml) rose water
- 8 egg whites
- A pinch of salt
- 1/6 cup (40 g) sugar

Custard

- 3 cups (750 ml) rose milk (milk and rose water)
- 8 egg yolks
- 2/3 cup (160 g) sugar

INGREDIENTS

- 1 1/4 to 2 lb (600 g to 1 kg) green papaya, cut into spaghetti strands
- 1/2 cup (125 ml) olive oil
- Juice of 2 lemons
- Salt and pepper
- 8 oz (240 g) blackberries
- 1 oz (30 g) chives, minced

- 1/3 cup (80 g) unsalted butter
- 2 1/2 oz (75 g) shallots, finely chopped
- 1 1/4 lb (600 g) monkfish medallions
- Salt and pepper
- 2/3 cup (160 ml) dry white wine
- Juice of one lemon
- 3/4 cup (150 g) carrots
- 1 cup (150 g) blackberries
- 1/4 cup (60 g) unsalted butter

Green Papaya Salad with Blackberries

4 servings • 2

Preparation: 15 Minutes

It is better to buy green papayas in speciality stores where you can have them cut into spaghetti strands. If you cannot find any, buy large papayas whole and slice thinly.

- About 2 hours before serving the salad, put the papaya into a bowl. Add the olive oil, lemon juice, salt, and pepper. Macerate it, stirring occasionally.
- When it is time to plate, add the blackberries and chives. Serve this dish chilled.

Monkfish Medallions with Carrot and Blackberry Juice

4 servings • Cooking time: 20 minutes • 3

Preparation: 30 Minutes

Medallions are pieces of fish 1 1/2" (4 cm) in diameter and 3/4" (2 cm) thick. Monkfish, also called headfish, has only one central bone. The taste is mild and it's ideal if you want to encourage children to eat fish.

- Grease a baking dish where the medallions won't be too tightly spaced nor too spread out.
- Sprinkle the shallots into the bottom of the dish and place the medallions on top. Season them with salt and pepper. Pour in the wine and lemon juice.
- Cook covered in the oven at 400 °F (200 °C) from 6 to 7 minutes until small white beads form on top of the medallions. Remove the medallions from the oven and keep warm. Reduce the cooking liquid by 80%.
- At the same time, juice the carrots and blackberries in a juicer. Pour the juice into the cooking liquid reduction and reduce again by half. Adjust the seasoning and finish with butter. Pour this over the medallions.
- Serve with rice or steamed potatoes.

Blackberry Liqueur

It is important that the fruit be very ripe.

- Using a juicer, remove all the fruit pulp, making sure to keep the seeds, which can be used to make vinegar. Strain the pulp through a cone-shaped or fine mesh sieve, then through cheese cloth. For every 4 cups (1 litre) of juice, add 3 cups (720 g) granulated sugar. Leave to dissolve for 1 hour.
- Add the eau-de-vie and vanilla. Let macerate for 3 weeks, stirring occasionally to dissolve the sugar.
- Strain and pour into opaque bottles. This liqueur can be aged for a few months.

INGREDIENTS

- 2 lb (1 kg) blackberries
- 8 oz (240 g) raspberries
- 8 oz (240 g) red currants
- 3 cups (720 g) granulated white sugar (for 4 cups /1 litre of juice)
- 6 cups (1.5 litres) eau-de-vie, 50%, or white rum
- 1/2 teaspoon vanilla extract

Blackberry Vinegar

- Put the seeds and the vinegar into a Mason jar and put the lid on tightly. Keep in a kitchen cupboard for 8 to 12 months. Strain before using.

- 10 oz (300 g) blackberry, raspberry, and red currant seeds
- 4 cups (1 litre) good quality red wine vinegar, acidified to 7% alcohol

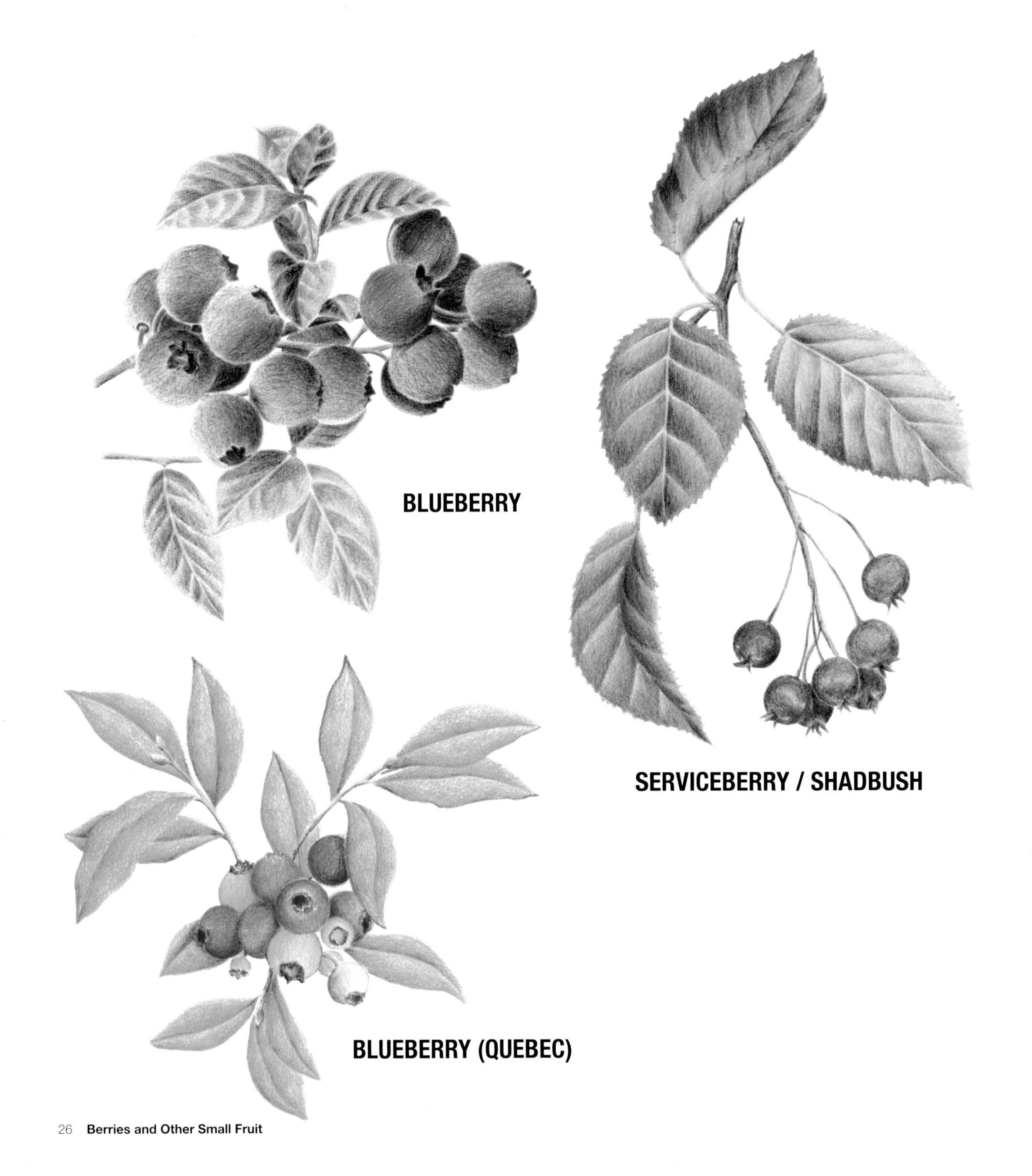
BLUEBERRY
SERVICEBERRY / SHADBUSH
BLUEBERRY (QUEBEC)

Blueberries | Canadian Blueberries | Serviceberries

Lowbush blueberry, *Vaccinium angustifolium*

Canadian blueberry, *Vaccinium myrtilloides*

Family: *Ericaceae*

Blueberry is a lovely Canadian word for what the French call "baie de l'airelle," or "myrtille" (*Vaccinium myrtillus*). The American blueberry differs from the European blueberry in the height of the bush and the size of the fruit. That said, the berries of both species are more flavourful in the wild where the blueberry is a shrub that does not exceed sixty centimetres in height. The leaves are twice as long as they are wide and they turn a magnificent gold or purple in the fall. The graceful bell-shaped flowers are white or pink. The fruit is bitter, refreshing, and antiseptic. What's more, it contains a pigment that promotes regeneration of retinal purple (necessary for the perception of light) and restores vitality to tired eyes. Blueberries and cranberries stand up well to freezing and drying.

Shadbush berries (or serviceberries) can be substituted for blueberries in any recipe. This shrub merits being grown as a hedge for its beauty when it blossoms. The shadbush produces an almost black fruit that is sweet, juicy, and delectable. The fruit makes excellent jellies, jams, and also a garnish for game meats, poultry, fish, and shellfish.

QUEBEC SHADBUSHES

Low shadbush, *Amelanchier humilis*
(Amélanchier bas)
Mountain juneberry, *Amelanchier bartramiana*
(Amélanchier de Bartram)
Allegheny serviceberry, *Amelanchier laevis*
(Amélanchier glabre)
Shadbush, *Amelanchier sanguinea*
(Amélanchier sanguin)
Running serviceberry, *Amelanchier stolonifera*
(Amélanchier stolonifère)

Cooking and Baking

Blueberries can accompany game birds and some delicatessen meats.

Chocolates, candies, jams, cakes, jellies, liqueur, fruit paste, vinegar.

Therapeutic Uses

Internal uses: arteriosclerosis, rheumatoid arthritis, diabetes, diarrhea, dysentery, enteritis, liver, hemorrhaging due to fragile capillaries, insufficient bile, menorrhagia (heavy menstrual bleeding), retinopathy, sequelae (conditions resulting from disease), uremia.

External Uses: freckles, eczema, thrush, pharyngitis, stomatitis (mouth inflammation).

Blueberry Potato Crêpes

4 servings • Cooking time: 10 minutes • 2

These crêpes can be served as a side with game meat dishes.

Preparation: 20 Minutes

- Cook the potatoes in salt water and purée them with a food mill. Add the milk to the hot purée. Cool. Next, add the flour and stir vigorously. Add the whole eggs and the egg whites. Season with salt and pepper. Using a spatula, fold in the cream to get the consistency of custard. Add the dried blueberries.
- Preheat the oven to 425 °F (220 °C) and heat the melted butter on a baking sheet. When it is very hot, spoon on small amounts of batter to make uniform circles. Put them in the oven and turn them over when the cooked side is a nice golden colour. Put the cooked crêpes on a plate and keep warm.
- While the crêpes are cooking, use a blender to emulsify 2 tablespoons of the boiling water with the lemon juice. Add the softened butter, little by little. Season to taste with salt and pepper. Place the crêpes on individual plates and spoon the lemon butter around them. Sprinkle with the chives.

INGREDIENTS

- 10 oz (300 g) Idaho potatoes, peeled
- 1 cup (250 ml) milk
- 1/2 cup (90 g) all-purpose flour
- 2 whole eggs
- 2 egg whites
- Salt and ground white pepper
- 6 tablespoons (90 ml) heavy cream (35%)
- 5 oz (150 g) dried blueberries
- 3/4 cup (180 g) unsalted clarified butter
- 2 tablespoons (30 ml) boiling water
- Juice of one lemon
- 1/2 cup (120 g) fresh butter, softened
- 2 oz (60 g) chives, minced

INGREDIENTS

- 12 veal grenadins (small veal scallops), 2 oz (60 g) each
- Salt and freshly ground pepper
- 2/3 cup (160 g) flour
- 1 cup (240 g) unsalted butter
- 1/3 cup (80 ml) cooking oil
- 2 dried shallots, finely chopped
- 6 tablespoons (90 ml) white wine
- 6 tablespoons (90 ml) cognac
- 3/4 cup (175 ml) thickened brown veal stock or demi-glace
- 1 cup (150 g) fresh blueberries

Blueberry Veal Grenadins

4 servings • Cooking time: 10 minutes • 2

Veal grenadin can be compared to beef tournedos. This milk-fed veal filet mignon is 1" (2.5 cm) long and 3/4" (2 cm) thick. The meat is very tender—in fact, the most tender part of the animal.

Preparation: 20 Minutes

- Put the grenadins onto paper towels. Season both sides uniformly with salt and pepper. Sprinkle the flour onto a plate. Heat half of the butter and oil in a heavy-bottom frying pan. Dredge each grenadin in flour and put them, one at a time, into the hot oil. Once they are seared, they will be a nice golden colour. Lower the temperature and cook them slowly. The internal temperature should reach 158 °F (70 °C).
- Remove the grenadins from the pan and keep them warm, but do not continue cooking. Remove the cooking fat and add the shallots, wine, and port to the pan. Reduce by half and add the veal stock. Simmer for 2 to 3 minutes. Finish the sauce with the remaining butter. Put the blueberries into the sauce. The temperature should not exceed 185 °F (85 °C), otherwise the fruit will burst. This dish goes extremely well with gratin dauphinois (a potato casserole).

Grilled Swordfish and Blueberry Hollandaise Sauce

4 servings • Cooking time: 15 minutes

Swordfish meat is tasty but dense and needs time to rest after cooking. Do not overcook it.

Preparation: 30 Minutes

- Wrap the swordfish well in paper towels. Set aside for at least 1 hour to remove as much moisture as possible. Season both sides with salt and pepper and brush lightly with oil.
- Sear the fish on each side on the hottest part of the grill then move the fish onto a part of the grill that is at medium heat. When the internal temperature reaches 158 °F (70 °C), move the fish to the coolest part of the grill to allow it to rest for 5 minutes.
- Prepare the hollandaise while the fish cooks. Using a coffee grinder, grind the dried blueberries and add them to the sauce.
- Serve the fish with steamed potatoes. The sauce should always be served on the side; otherwise the heat might cause it to separate.
- The barbecue or the grill should always have three heat zones: high, medium, and low heat.
- Nothing is more disappointing than grilled meat or fish with crisscross grill marks that are charred because the heat was too high.

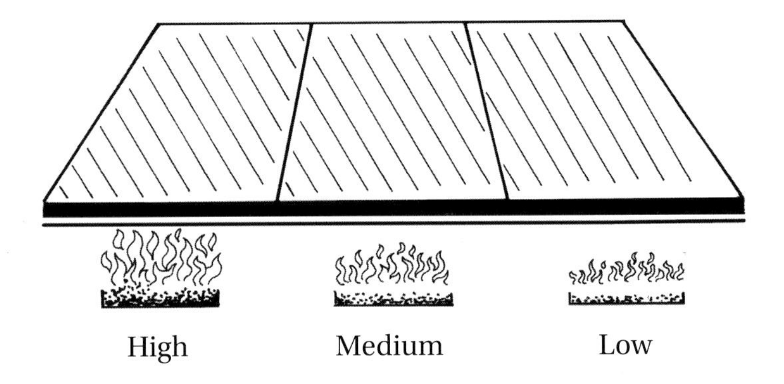

INGREDIENTS

- 4 swordfish steaks, 5 to 6 oz (150 to 180 g) each
- Sea salt and freshly ground pepper
- 6 tablespoons (90 ml) peanut oil
- 1 1/2 cups (375 ml) hollandaise sauce (see Basic Recipes)
- 4 oz (120 g) dried blueberries

Blueberry Chicken Supremes

4 servings • Cooking time: 30 to 40 minutes • 2

Supremes are chicken breasts that have the wing tip still attached.

Preparation: 30 Minutes

- Season the chicken supremes with salt and pepper. In a heavy-bottom pan, heat the oil and brown the chicken. Place the supremes in a large enough bowl to marinate and cool. Add the blueberry juice, wine, lemon juice, onions, bouquet garni, carrot, and bouillon. Cover with plastic wrap and leave at room temperature to marinate for a few hours.
- Pour the chicken and marinade into a Dutch oven and cook at 185 °F (85 °C), until the supremes reach an internal temperature of 167 °F (75 °C). Remove the supremes and all the accompanying vegetables from the Dutch oven. Cook the potatoes in the Dutch oven to thicken the liquid, adding the Veloutine to thicken it to the desired consistency. Adjust the seasoning and put the chicken supremes back in. Add the onions, celery, carrots, and blueberries. Simmer for a few minutes and serve in soup bowls. Top with the small croutons and chives.

INGREDIENTS

- 4 skinless chicken supremes (breasts), with wing tip
- Salt and freshly ground pepper
- 6 tablespoons (90 ml) cooking oil
- 3/4 cup (175 ml) blueberry juice
- 3/4 cup (175 ml) tannic red wine
- Juice of one lemon
- 12 small cipollini onions or 32 small, fresh pearl onions
- 1 bouquet garni
- 1 whole carrot
- 1 chicken bouillon cube or liquid, store bought
- 32 parisienne potatoes (potato balls)
- Veloutine (store-bought thickener)
- 1 celery stalk, cut into pieces
- 1 1/3 cups (200 g) blueberries
- 3 1/2 oz (100 g) small croutons
- 1 oz (30 g) chives, minced

Blueberry Syrup

Preparation: 30 Minutes

- Put the blueberries and water into a stockpot. Cook on high heat for 3 minutes; the fruit will burst on its own. Strain and allow the juice to ferment for 24 hours.
- Once more, gently strain the fermented juice through a sieve lined with cheesecloth. Weigh the amount of juice and add 2 cups (500 g) of sugar for every 1 lb (500 g) of juice. Mix well and heat until it just begins to boil. Cool completely and bottle.

INGREDIENTS

- 2 lb (1 kg) blueberries, very ripe and washed
- 12 cups (3 litres) distilled water
- 2 cups (500 g) granulated white sugar

INGREDIENTS

English Shortbread

- 3 egg yolks
- 1/2 cup (120 g) sugar
- 1 cup (180 g) flour
- 1/3 oz (10 g) baking powder, sifted
- 1/2 cup (120 g) salted butter, at room temperature

Fresh Goat Cheese Cream

- 4 oz (120 g) fresh goat cheese
- 1 oz (25 g) icing sugar
- 1/2 cup (125 ml) heavy cream (35%)

Nougatine (Praline)

- 1/3 cup (80 g) sugar
- 3 tablespoons (45 ml) water
- 7 oz (200 g) pistachios, shelled and toasted

Garnish

- 4 oz (120 g) blueberries, fresh or frozen
- 1/4 cup (60 g) honey

English Shortbread with Fresh Goat Cheese Cream and Nougatine

6 servings • Cooking time: 10 minutes • 3

Preparation: 50 Minutes

- In a mixer, beat the eggs and sugar until light in colour. Add the flour and baking powder. Add the butter and stop the mixer as soon as the dough is blended. Roll out the dough to a thickness of 2/3" (1.5 cm) between 2 pieces of parchment paper. Chill for 3 hours.
- Cut out the dough using a 2 1/2" (6 cm) cookie cutter. Bake the shortbread in round aluminum moulds in the oven for about 10 minutes at 300 °F (150 °C). Cool and turn out of the moulds.

Fresh Goat Cheese Cream

- In a mixer or with a whisk, blend the cheese and icing sugar on medium speed. Gradually add the cream and whip to soft peaks.

Nougatine

- In a pot, boil the sugar and water to make a light caramel. Add the pistachios and mix to coat them well. Spread them out on a baking sheet and cover with parchment paper. Cool before roughly chopping.

Garnish

- Put the cream cheese in the centre of the shortbread. In a saucepan, heat the honey and add the blueberries. Heat for a few seconds, long enough to soften the blueberries. Spoon the mixture around the shortbread and sprinkle with the nougatine.

Lemon Cream and Blueberry Macarons

4 servings • Cooking time: 10 minutes

Preparation: 40 Minutes

- Sift the ground almonds and the sugar.
- In a mixer, whip the egg whites on medium speed. When they have reached their maximum volume, add 1 tablespoon of sugar and whip for a few more seconds. The whites should hold together and be nice and shiny.
- Add the dry ingredients to the egg whites. Put the batter in a pastry bag and make macarons about 2" (5 cm) in diameter. Leave the macarons uncovered for about 30 minutes until a skin forms on them.
- Bake the macarons for 3 minutes at 350 °F (180 °C), until they have risen nicely. Lower the temperature to 325 °F (160 °C) and bake for 6 minutes more. Cool before removing them from the pan.

Lemon Cream

- Cook the lemon juice, eggs, sugar, and the butter in a saucepan while whisking. Continue to cook until the mixture begins to boil. Take it off the heat and add the gelatin.
- Strain the cream through a cone-shaped or mesh sieve. Cover tightly and chill.

Assembly

- Make "sandwiches" with the macarons, lemon cream, and blueberries.

INGREDIENTS

Macarons

- 5 1/2 oz (165 g) ground almonds
- 1 1/3 cups (330 g) icing sugar
- 5 egg whites

Lemon Cream

- 2/3 cup (160 ml) lemon juice
- 3 eggs
- 1/2 cup (120 g) sugar
- 3 oz (90 g) butter, diced
- 1 gelatin sheet

Assembly

- 4 oz (120 g) blueberries

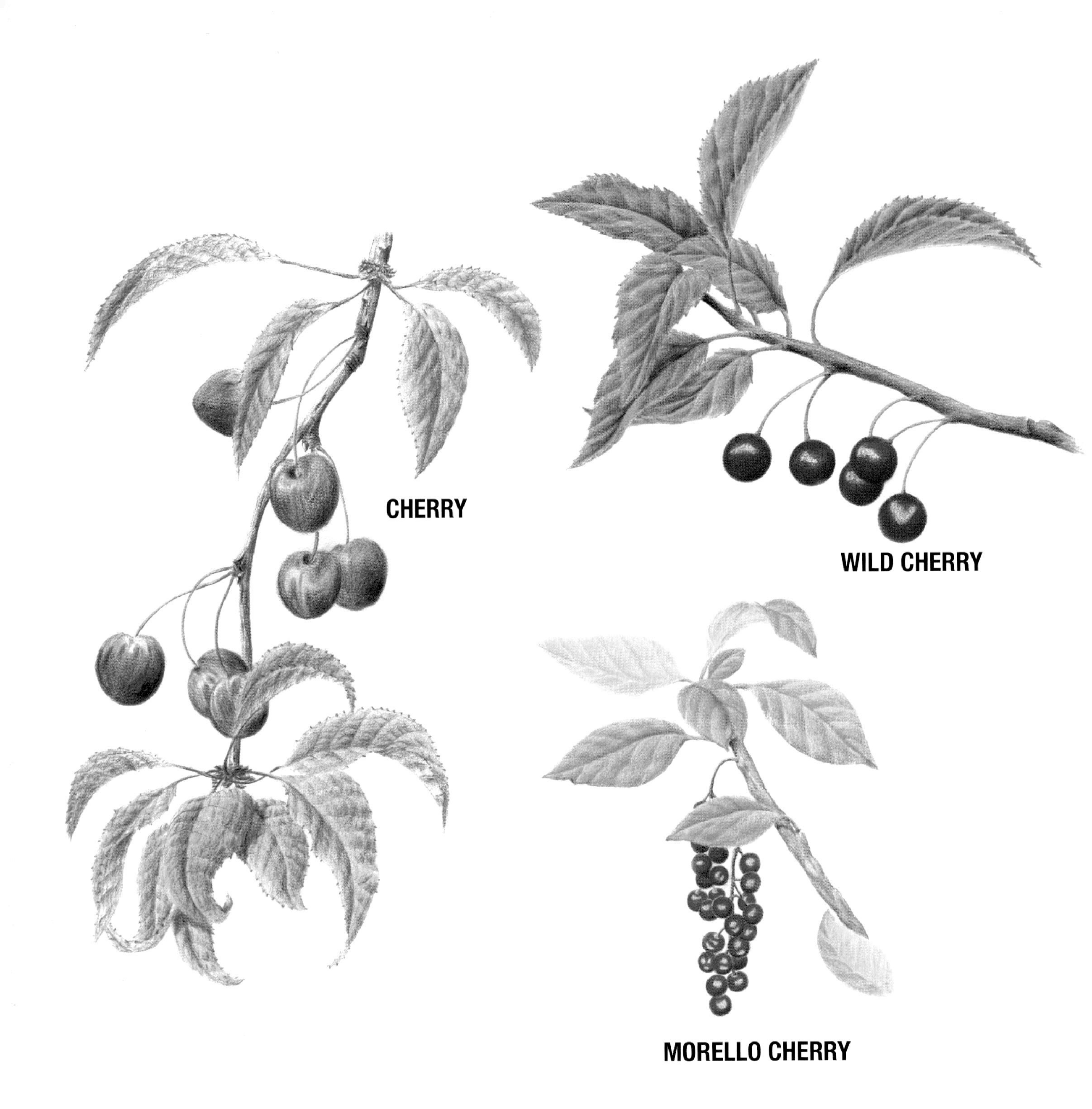
CHERRY
WILD CHERRY
MORELLO CHERRY

Cherries | Morello Cherries

Family: *Rosaceae*

We know of about six hundred varieties of cherries divided into four main groups: dark sweet, white sweet, tart, and duke cherries.

The trees come from two species: the sweet cherry *(Prunus avium)* and the sour (Morello) cherry *(Prunus cerasus)*.

The fruit of the dark sweet cherry tree looks like cherry hearts. The flesh is tender, sweet, and juicy, enjoyed as an eating fruit.

White Sweet or Bigaroon cherries (from the French word "bigarrer," meaning spotted—the cherries are often red and white) are firm, sweet, and crisp. The pit is red or yellowish red.

Duke cherries, which come from the hybridization of the wild cherry and cultivated cherry trees, are the eating fruit of choice, tender, very juicy, bright or dark red.

Tart or Morello cherries come from the smallest of the cherry trees. They have a short stalk, soft flesh that is very tart and a colourful juice. These small cherries, now grown in Quebec, are excellent for cooking. They are an ingredient in classic dishes such as Duck Montmorency.

Other Cherries

Pin cherry, *Prunus pensylvanica* (Cerisier de Pennsylvanie or petit merisier)
Black cherry, *Prunus serotina* (Cerisier tardif)
Chokecherry, *Prunus virginiana* (Cerisier à grappes)

Cooking and Baking

Duck, chicken, veal.
Compotes, jams, flans, cakes, jellies, sorbets, tarts.
Alcohol: cherry brandy , eau-de-vie, Guignolet (wild cherry liqueur), kirsch, Maraschino.

Therapeutic Uses

Anti-infective, antiarthritic, depurative (purifier), detoxifier, diuretic, demulcent (soothes & protects), immunity, laxative, minerals, sedative, tonic, hepatic and gastric disorders.

Sautéed Pork with Morello Cherries and Apples

4 servings • Cooking time: 30 minutes • 3

Preparation: 20 Minutes

- Peel the apples and dice them into 1/2" (1 cm) pieces. Pit the cherries. In a sauté pan, heat the butter and simmer the apples and cherries at 185 °F (85 °C). The cherries should cook without bursting.
- Peel the onions and prepare the bouquet garni. Heat the cooking oil in a cast iron or heavy-bottom frying pan. Quickly brown the diced pork. Season with salt and pepper, pour in the wine, and add the tomato paste. Cook to remove the acidity and add the veal stock. Add the onions, bouquet garni, and garlic. Cook slowly until the cubes of meat are fork tender. Strain the sauce and thicken with the white roux. Cook for 10 minutes and strain again. Put the cubes of pork back into the sauce with the onions, cherries, and apples. Simmer for 5 minutes and pour in the kirsch.
- Serve hot with steamed potatoes.

INGREDIENTS

- 4 Russet apples or other cooking apples
- 9 oz (270 g) Morello cherries, fresh or canned
- 1/2 cup (120 g) unsalted butter
- 24 small white pearl onions
- 1 bouquet garni
- 4 cloves of garlic, unpeeled
- 6 tablespoons (90 ml) cooking oil
- 1 1/4 lb (600 g) pork tenderloin, cut into 3/4" (2 cm) cubes
- Salt and freshly ground pepper
- 3/4 cup (175 ml) white wine
- 1/4 cup (60 ml) tomato paste
- 2 cups (500 ml) unthickened veal stock
- White roux
- 5 tablespoons (75 ml) kirsch

INGREDIENTS

- 4 Northern pike, 1 to 1 1/4 lb (480 to 600 g) each
- 6 tablespoons (90ml) hazelnut oil or another nut oil
- 10 oz (300 g) sea beans (salicornia)
- 3 dried shallots, finely chopped
- Salt and freshly ground pepper
- 3/4 cup (175 ml) Morello cherry juice
- 3/4 cup (175 ml) white Noilly Prat vermouth
- 1 cup (250 ml) thickened lobster sauce, homemade or store bought
- 5 oz (150 g) Morello cherries, pitted
- Juice of one lemon
- 3 1/2 oz (100 g) knob of butter

Northern Pike with Sea Beans, Lobster Sauce, and Cherry Juice

4 servings • Cooking time: 25 minutes • 3

It is better to make this dish with Northern pike because the small bones dissolve into the meat. Use only the centre cut, the thickest part of the pike. Scale it well. Wash and dry the fish and keep it in the fridge. Taste the sea beans raw. If they are too salty, blanch them once or twice.

Preparation: 40 Minutes

- Choose a baking dish where the pike will fit snugly. Pour in the hazelnut oil, sprinkle with the shallots, and top with the sea beans. Season the pike with salt and pepper. Place it on top of the sea beans, belly down. Pour in the Morello cherry juice and Noilly Prat. Cover with a piece of parchment paper and bake at 350 °F (180 °C) until it reaches 158 °F (70 °C) close to the bone.
- While the fish cooks, heat the lobster sauce and then add the cooking liquid from the pike. Reduce to your desired consistency. Add the Morello cherries and the lemon juice. Finish the sauce with butter and adjust the seasoning.
- Put some sea beans onto the bottom of each plate, then top with the fish. Sauce the plate.
- Serve hot with steamed potatoes.

Duck Montmorency

4 servings • Cooking time: 30 minutes • 3

The "magret" is the lean fillet of a goose or duck that produces fois gras. Fillets from Mulard or Muscovy ducks are generally more tender because they are sedentary birds.

INGREDIENTS

- 2 duck fillets ("magrets"), 12 oz to 1 lb (360 g to 480 g)
- Salt and freshly ground pepper
- 5 tablespoons (75 ml) kirsch
- 1 cup (250 ml) Morello sauce (see Basic Recipes)
- 1/2 cup (120 g) unsalted butter
- 7 oz (200 g) Morello cherries, pitted

Preparation: 40 Minutes

- Generously season the fillets with salt and pepper. Heat a frying pan or a non-stick sauté pan and begin to cook the skin side of the fillets to render a bit of the fat. Reduce the heat and turn over the fillets. Cook slowly so that the skin is crispy. To cook until just done, the internal temperature should reach 129 °F (54 °C). Remove the excess cooking fat and flambé with the kirsch.
- Remove the fillets. Keep warm on a rack to rest them. In a frying pan or a sauté pan, heat the Morello sauce, finish with butter, and add the cherries. Adjust the seasoning.
- Pour the sauce onto the bottom of the plates and thinly slice the fillets. Serve with crosnes (Chinese artichoke) or salsify (oyster plant).

Cherry Syrup

A few drops of this syrup in citrus-flavoured soda pop makes a Cherry Diabolo. A diabolo (a sugary non-alcoholic drink) can also be flavoured with mint, grenadine, and blackcurrant.

- 2 lb (1 kg) sour cherries (Morellos, etc.)
- 3 1/2 cups (850 g) granulated white sugar

Preparation: 40 Minutes

- Wash the cherries well. Grind them with the pits to make a purée. The pits should be thoroughly ground. Ferment for 35 to 40 hours in a large-mouth Mason jar. Strain, then strain again through cheesecloth until the juice is clear.
- Measure the juice and add the granulated sugar (3 1/2 cups/850 g for every 2 1/2 cups/625 ml of juice). Bring to a boil for 1 or 2 minutes. Skim. Strain again and pour into sterilized bottles.

INGREDIENTS

Mixture

- 2 1/2 oz (75 g) butter, softened at room temperature
- 1/2 cup (120 g) sugar
- 1 egg
- 1 egg yolk
- 3 1/2 oz (100 g) ground almonds
- 1 tablespoon (45 g) cornstarch
- 6 tablespoons (90 ml) heavy cream (35%)

Vanilla Ice Cream

- 2 cups (500 ml) milk
- 1 cup (250 ml) cream
- 1/2 cup (120 g) sugar
- 1 vanilla bean
- 6 egg yolks

Garnish

- 10 oz (300 g) Morello cherries, halved and pitted

Cherry Clafoutis and Vanilla Ice Cream

6 servings • Cooking time: 30 minutes

Preparation: 40 Minutes

- In a mixer, cream the butter and sugar. Add the egg and egg yolk. Add the dry ingredients and the cream, alternating the two. Chill the batter for 48 hours.

Vanilla Ice Cream

- Boil the cream, milk, and half the sugar with the vanilla bean, split and with the seeds scraped out. Remove from the heat and let the flavours infuse for 10 minutes. Beat the egg yolks with the rest of the sugar. Boil the milk and cream again. While stirring constantly, gradually pour the hot liquid over the yolks to warm them but not cook them. Put the mixture back into the saucepan on medium heat. Cook while stirring constantly, until the cream coats the back of a spoon. Chill over ice and keep in the fridge overnight. The next day, put the ice cream into an ice cream maker and process.

Garnish

- Pour the clafoutis mixture into greased and floured pans to three-quarters full. Add the Morellos. Bake at 350 °F (180 °C) for about 12 minutes, depending on the size of the clafoutis. Cool before unmoulding. Serve with the vanilla ice cream.

Spiced French Toast, Cherry Jam, and Chocolate Curls

4 servings • Cooking time: 45 minutes • 3

If you cannot find Montmorency cherries, you can use bing or rainier cherries as a substitute, but only use 3 1/2 oz (100 g) of sugar.

Preparation: 30 Minutes

- In a saucepan on medium heat, cook the cherries and sugar. After about 45 minutes, when the cherries appear cooked and the liquid is syrupy, strain through a cone-shaped or mesh sieve. Put the syrup back on to boil. Add the pectin and sugar and cook for about 1 minute. Put the cherries back in and cook for another minute. Chill.

French Toast Mixture

- In a bowl, mix together well the sugar, lemon and orange zest, and the spices. Add the eggs and cream. Finish by mixing in the milk.

Assembly

- Soak the bread in the French toast mixture. In a non-stick frying pan on medium heat, brown the bread on all sides until nicely golden. Finish cooking for 2 minutes in the oven at 350 °F (180 °C). Serve the French toast with the jam and chocolate curls.

INGREDIENTS

Cherry Jam

- 13 oz (400 g) Montmorency cherries, pitted
- 7 oz (200 g) sugar
- 3 g pectin (the tip of a knife)
- 3 g sugar (the tip of a knife)

French Toast Mixture

- 1/2 cup (120 g) sugar
- Zest of a lemon
- Zest of an orange
- 1 teaspoon (5 g) cinnamon
- 1/2 vanilla bean
- 1/2 teaspoon (2.5 g) each: ground cardamom, nutmeg, star anise, pepper
- 5 eggs
- 2 cups (500 ml) milk

Assembly

- Brioche bread, sliced
- Block of chocolate, shaved

Cloudberries

Rubus chamaemorus (Plaquebières)
Other names: Bakeapple, yellowberry
Family: *Rosaceae*

Laplanders call the cloudberry "marsh raspberry," and they make a liqueur with it. In Quebec, cloudberry is very well known on the North Shore, and the Montagnais call it "chicouté," which in their language means fire—most likely because of the colour of the ripe fruit.

Some people use the old French name "plat-de-bièvre," or beaver food. Cloudberries are distinctive from other plants from the genus *Rubus* because the shoots take three years to produce leaves. The ripe fruit is a beautiful amber colour, but it is not sweet and must be eaten chilled or preserved. They are excellent in sorbets, cakes, and mousses.

Cooking and Baking
Game meat and fish sauces
Jams, cakes, jellies, fruit pastes, sorbets.

Therapeutic Uses
Anhidrosis (absence of perspiration), anxiety, asthenia (weakness), constipation, dermatosis (skin condition), dyspepsia, gastro-intestinal difficulties, fever, gout, rheumatism.

Frog's Legs Soup with Cloudberry Juice

4 servings • Cooking time: 30 minutes • 3

Preparation: 30 Minutes

- Put the onions, carrots, and celery into a stock pot. Pour in the water. Add the bouquet garni, peppercorns, lemon juice, and wine. Salt lightly.
- Cook 20 to 25 minutes to make a tasty court-bouillon. Add the frog's legs and simmer for 25 minutes (194 °F / 90 °C). Take out the frog's legs and remove all the meat from the bones. Cover with a wet towel so the meat doesn't dry out. Put all the small bones into the stock pot and simmer for another15 minutes. Strain through a cone-shaped or mesh sieve and reduce by half. Thicken slightly with the white roux. Add the cream and simmer gently to your desired consistency.
- Cut the unpeeled zucchinis in a small dice. Adjust the seasoning of the stock. Add the cloudberry juice.
- Place the frog meat and zucchini in the bottom of soup bowls. Pour the hot creamy stock over everything. Serve immediately.

* 9/12: 9 to 12 frogs per 1 lb (480 g)

INGREDIENTS

- 1 Spanish onion, minced
- 1 carrot, minced
- 1 celery stalk, minced
- 8 cups (2 litres) water
- 1 bouquet garni
- 12 black peppercorns
- Juice of one lemon
- 3/4 cup (175 ml) white wine
- Salt
- 1 1/4 cups (600 g) frog's legs (9/12*)
- White roux
- 1 cup (250 ml) heavy cream (35%)
- 2 zucchinis
- 2/3 cup (160 ml) cloudberry juice or light apple juice
- Salt and ground white pepper

Sautéed Chicken Livers Scented with Cloudberries

4 servings • Cooking time: 8 minutes • 3

INGREDIENTS

- 1 lb (480 g) chicken livers
- 2 dried shallots, finely chopped
- 1 cup (250 ml) cloudberry juice or pear juice
- 2/3 cup (160 ml) sherry
- 3/4 cup (175 ml) unthickened brown chicken stock, homemade or store bought
- 2 tablespoons (30 ml) olive oil
- 1 cup (200 g) Chinese cabbage, minced
- 2 cloves of garlic, finely chopped
- 5 tablespoons (75 ml) sherry vinegar
- Salt and freshly ground pepper
- 3 1/2 oz (100 g) small croutons

Preparation: 15 Minutes

- Clean the chicken livers well, removing all the veins. Place them in a saucepan and cover with cold water. Bring to a boil, skim, and cook for 12 minutes at 185 °F (85 °C). Cool under a stream of cold water. Drain and set aside.
- In a pot, mix the shallots, cloudberry juice, and sherry. Simmer and reduce by 80%. Add the chicken stock. Simmer gently for 20 minutes.
- Heat the olive oil in a large frying pan and sauté the cabbage with the garlic and sherry vinegar. Cook for 3 to 4 minutes. Carefully mince the livers. Season with salt and pepper.
- Add the cloudberry juice reduction to the cabbage. Five minutes before serving, add the pieces of chicken liver. Serve hot in soup bowls. Sprinkle with the small croutons.

INGREDIENTS

- 1 cup (250 ml) milk
- 1 cup (250 ml) heavy cream (35%)
- 1/3 cup (80 g) sugar
- 1 teaspoon (5 g) cinnamon, ground
- 5 egg yolks
- Cloudberry jam

Cloudberry and Cinnamon Crème Brulée

4 servings • Cooking time: 40 minutes • 3

Preparation: 30 Minutes

- In a saucepan, boil the milk, cream, and half the sugar. Add the cinnamon and leave to infuse for 5 minutes. Beat the egg yolks with the remainder of the sugar. Gradually pour the hot liquid onto the egg yolks, while stirring. Strain.
- Cover the bottom of some ramekins with a thin layer of cloudberry jam. Pour in the crème brulée mixture on top of the jam. Put them into the oven in a baking dish with hot water halfway up the ramekins. Bake for about 40 minutes at 300 °F (150 °C), until the cream mixture no longer jiggles. Cool and then put them into the fridge for a few hours.
- Sprinkle the crème brulée with a thin layer of sugar and caramelize using a blow torch.

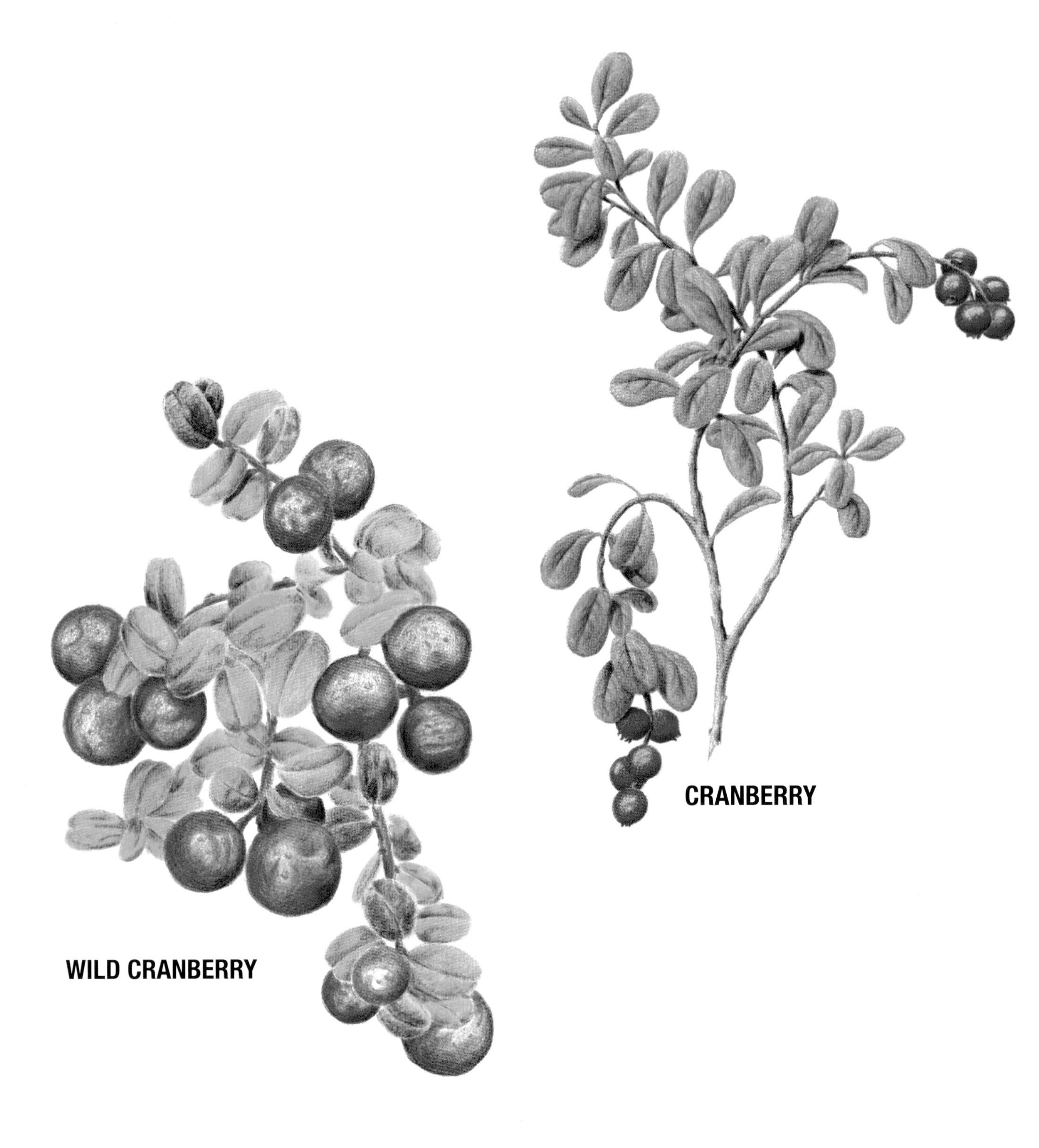
CRANBERRY
WILD CRANBERRY

Cranberries

Family: *Ericaceae*

Cranberries are small, spherical, floury, and tart berries that turn red when ripe, generally in September. They blossom in spring (cowberry) or the beginning of summer (small and large cranberries). The bell flowers are white or pink. Common Nordic varieties grow in Quebec from île d'Orléans to the tundra.

Small cranberry is a dwarf shrub. It is smaller and more common than the large variety. The fruit of the small and large cranberry plants is very similar and used in the same way.

Cranberries are part of the same family as blueberries, but cranberries are much more tart. They are an oval, round berry that is twice as large as blueberries. Other members of the *Ericaceaes* family from South America include: gualicon (*Macleania pentaptera*), which resembles an elongated blueberry; taglli (*Pernettya parvifolia*), which grows in desert areas and under dried grass.

Other Varietes

Bog blueberry, *Vaccinium uliginosum*
(Airelle des marais)
Highbush blueberry, *Vaccinium corymbosum*
(Airelle en corymbe)
Dwarf bilberry, *Vaccinium cespitosum*
(Airelle gazonnante)
Cowberry; lingonberry, *Vaccinium vitis-idaea*
(Airelle vigne d'Ida)
Small cranberry, *Vaccinium oxycoccos*
(Petit atoca; canneberge)
Large cranberry, *Vaccinium macrocarpon*
(Gros atoca; canneberge)

Cooking and Baking

Garnishes for game meats and fish.

Compotes, jams, cakes, jellies, ice cream, sorbet.

Therapeutic Uses

Internal uses: arteriosclerosis, rheumatoid arthritis, diabetes, diarrhea, dysentery, enteritis, liver, hemorrhaging due to fragile capillaries, insufficient bile, menorrhagia (heavy menstrual bleeding), retinopathy, sequelae (condition due to illness), uremia.

External uses: freckles, eczema, thrush, pharyngitis (sore throat), stomatitis.

Eggs in Coconut Milk with Dried Cranberry

4 servings • Cooking time: 8 minutes • 2

Fresh turmeric is an orange rhizome from the ginger family. It has a very pronounced taste and is found in Asian supermarkets. It can be replaced with ground turmeric.

Preparation: 15 Minutes

- Peel and mince 5 of the shallots. Peel the garlic and remove the sprout. Cut the chilies in half and remove the seeds with a knife to avoid contact with your fingers. In a blender, process the shallots, garlic, chilies, ginger, and turmeric on high to obtain a spicy paste.
- In a saucepan, bring the coconut milk to a boil. Add the spice paste, the dash of turmeric, and the eggs. Simmer for 5 minutes. Set aside.
- Peel the 8 other shallots and cook in salted water to *al dente*. Drain them well on paper towels and sear them in the oil at 455 °F (235 °C). Drain on a paper towel while keeping them warm.
- Grind the cranberries in a coffee grinder. To finish the dish, add the tamarind juice, the shallots, adjust the seasoning, and sprinkle with the ground cranberries.

INGREDIENTS

- 13 small dried shallots
- 3 cloves of garlic
- 5 Mexican chilies, hot
- 1 oz (30 g) fresh ginger, peeled and minced
- 1/4 teaspoon fresh turmeric, peeled and minced
- 2 cups (500 ml) coconut milk
- A dash of ground turmeric
- 8 hard-boiled eggs
- Oil for frying
- 3 1/2 oz (100 g) dried cranberries, unsweetened
- 1 tablespoon (15 ml) tamarind juice
- Salt

Caribou Chops with Cranberries

4 servings • Cooking time: 10 minutes • 2

Caribou merits a higher place in the culinary order because it is one of the best game meats in the world. Its food value is broad and the caribou's rapid circulatory system gives the meat a distinctive character. If the animal is young, you don't even need a knife to cut the main muscles. Do not marinate it unless the animal is older.

Preparation: 20 Minutes

- Season the caribou chops with salt and pepper. In a heavy-bottom or cast-iron frying pan, heat the oil and butter. Sear the chops on each side to brown them. Reduce the heat to get the desired temperature: rare, 125 °F (52 °C); medium 133 °F (56 °C); well done, more than 140 °F (60 °C). It is not recommended to serve this cut of meat well done or it will be tough. Remove the chops and keep warm.
- Remove the excess cooking fat and put the shallots into the pan. Deglaze with red wine. Add the sauce *grand veneur*, cranberry liqueur, and the cranberries. Adjust the seasoning and simmer until the temperature of the sauce exceeds 185 °F (85 °C) or the fruit bursts.
- Place the chops in the bottom of the plate and sauce. Cattail hearts, crosnes (Chinese artichoke), chestnut, and celery root purée all make excellent accompaniments.

INGREDIENTS

- 4 caribou chops (if the animal is old), bone in, 6 oz (180 g) each or 8 caribou chops (if the animal is young), bone in, 3 oz (90 g) each
- Salt and freshly ground pepper
- 6 tablespoons (90 ml) cooking oil
- 1/2 cup (120 g) unsalted butter
- 3 dried shallots, finely chopped
- 3/4 cup (175 ml) full-bodied red wine
- 1 cup (250 ml) sauce *grand veneur* (sauce for game meat—see Basic Recipes)
- 1/3 cup (80 ml) cranberry liqueur (recipe follows)
- 5 oz (150 g) fresh cranberries

INGREDIENTS

- 3 lb 5 oz (1.5 kg) fresh cranberries
- 7 oz (200 g) ripe raspberries
- 3 1/2 oz (100 g) cranberry leaves (optional)
- 1 cinnamon stick
- 1 clove
- 4 cups (1 litre) white rum, 40%
- 2 cups (480 g) granulated white sugar
- 2/3 cup (160 ml) distilled water

Cranberry Liqueur

Preparation: 20 Minutes

- Wash and drain the cranberries well. Put the cranberries, raspberries, cranberry leaves, cinnamon, and clove into a wide-mouthed Mason jar. Add the white rum. Cover with an airtight lid and macerate for at least 45 days. Cellar it.
- Drain the fruit after macerating without pressing it, to collect a clear liquid. Mix the sugar and water together and bring to a boil. Cool and carefully mix in the cranberry juice. Pour into opaque bottles. Cork tightly and wait at least 2 months before drinking.

- 1/2 cup (125 g) sugar
- 3 1/3 oz (100 g) pecans, toasted
- 2 egg yolks
- 1/2 teaspoon (3 g) cardamom
- 1/2 cup (125 ml) mascarpone cheese
- 2 egg whites
- 1 cup (250 ml) heavy cream (35%)
- 1/2 cup (75 g) dried cranberries
- 1 cup (150 g) fresh cranberries
- 1/4 cup (60 g) honey

Cranberry and Pecan Nougat Confit

4 servings • Cooking time: 10 minutes

Preparation: 30 Minutes

- Put 1/4 cup (60 g) of the sugar in a small saucepan. Cover with water and cook until it caramelizes. Add the pecans, coat them well with the caramel, and spread them out on a parchment paper-lined cookie sheet. Chill and chop the nougat.
- In a mixer, beat the egg yolks, 25 g of sugar and the cardamom. Blend the beaten yolks into the mascarpone. Next, whip the egg whites and fix with the remainder of the sugar. Lastly, whip the cream to soft peaks (not too firm). Fold the whites and cream, alternately, into the mascarpone mixture. Fold the nougat and the dried cranberries into the mixture and put into a mould lined with plastic wrap. Keep in the freezer for 24 hours.

Assembly

- Heat the honey and add the fresh cranberries. Cook until they burst and make a jam. Take the nougat out of the freezer 2 minutes before serving. Cut the nougat and serve it with the hot cranberries.

INGREDIENTS

- 4 eggs
- 3/4 cup (180 g) sugar
- 5 oz (150 g) dark chocolate
- 1 cup (240 g) butter
- 1/2 cup (120 g) flour, sifted
- 2 1/2 oz (75 g) hazelnuts, roughly chopped
- CHOCOLATE CREAM
- 3/4 cup (175 ml) heavy cream (35%)
- 5 oz (150 g) milk chocolate

Garnish

- 1/4 cup (60 g) honey
- 2/3 cup (100 g) cranberries, fresh or frozen

Chocolate Hazelnut Cookies with Chocolate Cream and Pan-fried Cranberries

4 servings • Cooking time: 10 minutes

Preparation: 30 Minutes

- In a mixer, cream the eggs and sugar until light in colour. Melt the chocolate and the butter in the microwave in 30-second intervals. Lower the mixer speed and blend the chocolate and butter into the eggs; add the flour and hazelnuts.
- Bake for 10 minutes at 350 °F (180 °C) in a buttered pan lined with parchment paper.

Chocolate Cream

- Boil the cream and pour it over the chocolate in a mixing bowl. Beat the chocolate cream with a whisk. Cover the bowl with plastic wrap and put it in the fridge for at least 2 hours.
- To serve, lightly whip the chocolate cream, making sure it doesn't become grainy.

Assembly

- Garnish each cookie with the chocolate cream.
- In a pan, heat the honey and pan-fry the cranberries until they burst. Serve the hot cranberries with the cookie.

RED CURRANT
GOOSEBERRY
BLACK CURRANT

Currants

Family: *Saxifragales*

Other Varieties

Gooseberry, *Ribes grossularia*
(Groseillier cultivé)
Prickly wild gooseberry, *Ribes cynosbati*
(Groseillier des chiens)
Hairy gooseberry, *Ribes hirtellum*
(Groseillier hérissé)
White currant, *Ribes sativum*
(Groseillier rouge ou gadellier rouge)
Black currant, *Ribes nigrum* (Cassis)

Cooking and Baking

Red currants go very well with game meat and they are excellent in jams, cakes, jellies, juice, liqueurs, fruit paste, and pies.

Therapeutic Uses

Anxiety, arthritis, dyspepsia, liver, gout, digestive and urinary inflammations, rheumatism.

Cultivated red currants actually come from wild gooseberries. The red currant shrub can grow to 1 to 1.5 metres high. In July and August, fruit forms in lovely bunches with brilliant berries: red, pink, and a vivid yellow. Currants are quite delicate and need to be harvested at just the right moment. Picked too early, they are sour; too late, they bolt to seed. Currants were first cultivated in Holland in the Middle Ages. Their French name, groseille, comes from the Middle Dutch word *croesel.* There are many kinds of currants in Quebec.

Kir is an aperitif that is named for Canon Kir, a former deputy mayor of Dijon. It is a mix of the white grape aligoté from Burgundy and crème de cassis. The black currant berry, which is readily found in Quebec, is a small fruit on a shrub that can reach 1.3 metres in height. The pulp contains lots of seeds. It is slightly tart, juicy, and delicious. It is used in pastries, chocolate making, and to make crème de cassis and fruit paste. Black currants are also good in sauces for game meat, some poultry, and fish.

Boar Chops with Red Currants

4 servings • Cooking time: 15 minutes • 3

Boar should be marinated for two reasons. In Quebec, boar is a domestic animal and a marinade will improve the flavour of an already excellent quality meat. If the chops are from an older animal, the marinade will tenderize the meat.

Preparation: 30 Minutes

- Place the boar chops on a plate. Sprinkle with the onions, carrots, and celery. Add the bay leaf, thyme, juniper berries, black pepper, wine, vinegar, olive oil, salt, and pepper. Cover with plastic wrap and marinate for 24 to 36 hours.
- Take the chops out of the marinade and reduce the marinade by 80%. Add the sauce *grand veneur* and simmer for 15 minutes. Strain. Adjust the seasoning and keep warm. Season the boar chops with salt and pepper. Heat the butter in a heavy-bottom frying pan and sear the chops on each side. Lower the heat when they have reached an internal temperature of 175 °F (80 °C). Four or 5 minutes later, remove the chops and set aside. Remove the excess fat from the pan. Add the sauce and the red currants. Heat but do not exceed 185 °F (85 °C), because the currants should not burst.
- Plate the boar chops and cover with the sauce. Serve with celery root purée.

INGREDIENTS

- 4 boar chops, 5 to 6 oz (150 to 180 g) each
- 1 onion, minced
- 1 carrot, thinly sliced
- 1 celery stalk, thinly sliced
- 1/2 bay leaf
- 1 sprig of thyme
- 8 juniper berries
- 8 black peppercorns
- 1 1/4 cups (300 ml) tannic red wine
- 1/4 cup (60 ml) red wine vinegar
- 1/4 cup (60 ml) olive oil
- Salt and pepper
- 1 cup (250 ml) sauce *grand veneur* (see Basic Recipes)
- 2/3 cup (160 g) salted butter
- 1/4 cup (60 ml) olive oil
- 5 oz (150 g) red currants

INGREDIENTS

- 3 eggs
- 1/2 cup (120 g) sugar
- 1 cup (250 ml) milk
- 1 cup (250 ml) heavy cream (35%)
- Ground nutmeg
- 2 1/2 cups (375 g) brioche bread, cubed
- 1/2 cup (75 g) black currants

- 9 oz (270 g) ripe black currants
- 6 cups (1.5 litres) white rum or eau-de-vie, 45%
- A pinch of cinnamon
- 1/2 a vanilla bean
- 2 egg whites
- 1 lb (480 g) white granulated sugar
- 6 tablespoons (90 ml) boiling water

Black Currant and Nutmeg Brioche Pudding

4 servings • Cooking time: 15 to 20 minutes

Preparation: 20 Minutes

- Cream the eggs and sugar with a whisk. Add the milk, cream, and nutmeg. Mix in the cubed bread and leave in the fridge for 1 hour. Add the black currants. Pour into individual non-stick baking dishes and place these in a metal pan. Pour in water to halfway up the baking dishes. Bake in the oven for 15 to 20 minutes at 350 °F (180°C). Serve warm.

Black Currant Liqueur

Preparation: 20 Minutes

- Put the black currants into a large-mouth Mason jar with an airtight stopper and cover with rum. Add the cinnamon and vanilla bean. Macerate for 15 days. Drain the berries and press them to collect the juices.
- Decant and strain through a coffee filter. If the liquid is still cloudy, clarify it with an egg white. Shake well. Let settle for 24 hours. Decant and strain again.
- Dissolve the sugar into the 6 tablespoons (90 ml) of boiling water. Cool. Add the syrup and the black currant juice to the macerating liquid. If necessary, top up with more white rum or eau-de-vie to obtain 4 cups (1 litre) of liqueur. Shake well. Cellar it for a few months.

Saddle of Rabbit with Black Currants

4 servings • Cooking time: 40 minutes • 3

Ask your butcher to debone the rabbit saddles and to coarsely chop the bones.

INGREDIENTS

- 2 large rabbit saddles
- Salt and freshly ground pepper
- 2 lb (1 kg) spinach, rinsed well
- 3 1/2 oz (100 g) hazelnuts, toasted and crushed
- 1 egg white
- A pinch of ground nutmeg
- 1/4 cup (60 ml) olive oil
- 1 cup (240 g) unsalted butter
- 1 onion
- 1 carrot
- 1 celery stalk
- 1 sprig of thyme
- 1/2 bay leaf
- 3/4 cup (175 ml) dry white wine
- 6 tablespoons (90 ml) crème de cassis
- 3/4 cup (175 ml) brown rabbit stock or store-bought brown veal stock
- 7 oz (200 g) ripe black currants

Preparation: 30 Minutes

- Season the inside of the saddles with salt and pepper. Set aside in the fridge.
- In a stock pot, boil 4 cups (1 litre) of water with 3/4 teaspoon of salt. Toss the spinach into the stock pot and cook for 5 to 8 minutes. Rinse in cold water, drain, and wring out the spinach, squeezing it with your hands. Chop the spinach and mix it in a bowl with the hazelnuts and the egg white. Season with salt and pepper and add the nutmeg.
- Spread the spinach mixture onto the saddles and season with salt and pepper. Press the mixture all around the inside of the saddles. Close with the meat side on the bottom.
- Heat the oil with half the butter in a sauté pan. Brown the saddles on all sides. Place in the oven at 400 °F (200 °C), basting often. Chop the onion, carrot, and celery. Halfway through cooking (internal temperature of 113 °F / 45 °C), add the vegetables, thyme, and bay leaf. Cook again until the internal temperature is 162 °F (72 °C). Remove the saddles and add the chunks of bone. Brown them with the vegetables. Remove the excess fat and add the wine and crème de cassis. Simmer to evaporate the alcohol, then pour in the rabbit stock. Cook for 20 minutes and strain. Add the black currants. Finish the sauce with the butter and keep the temperature at 175 °F (80 °C) to ensure the berries don't burst.
- Pour the sauce into the bottom of the plates. Thinly slice the saddles and arrange the meat in the sauce. Delicious with French beans.

Goose Breast with Red Currants

4 servings • Cooking time: 60 minutes or more • 3

Wild goose should be prepared according to its age. It will not need marinating if it's young, but an older goose will have to be marinated. Generally, you can tell by pushing the tip of your index finger into the breast. If there is some resistance, it's a young bird. The harder the breast is, the older the goose.

INGREDIENTS

- 2 goose breasts, farmed or wild, 1 to 1 1/4 lb (480 to 600 g) each
- 5 to 6 strips of frozen bacon fat or back fat, 1/2" wide and 1/2" thick (1 cm x 1 cm), the same length as the breasts
- 6 tablespoons (90 ml) cooking oil
- 10 oz (300 g) mirepoix (onion, carrot, and celery), cut in large pieces
- 1 cup (250 ml) dry white wine
- 3/4 cup (175 ml) red currant juice
- 1 bouquet garni
- Salt and freshly ground pepper
- 2 cups (500 ml) unthickened veal stock, homemade or store bought
- White roux or an equivalent thickener
- 7 oz (200 g) red currants

Preparation: 30 Minutes

- Wrap the strips of bacon fat or back fat around each breast lengthwise. In a heavy-bottom frying pan, heat the oil and sear the breasts. Place them in a container with the mirepoix, wine, red currant juice, bouquet garni, salt, and pepper. If the goose is older, marinate for 24 hours. Otherwise, skip this step.
- Put all the ingredients into a small Dutch oven. Add the poultry stock. Season it with salt and pepper and cook at 194 °F (90 °C), until it is fork tender. Keep warm. Strain the cooking liquid and thicken with the roux to the consistency of your liking. A few minutes before serving, add the red currants to the sauce. Simmer for a few minutes without bringing it to a boil so the berries don't burst.
- Sauce the bottom of the plates and slice the breasts. Arrange the meat on the plates and cover with more sauce. This dish is very good with wild rice.

INGREDIENTS

- 4 partridges, farmed or wild
- Salt and freshly ground pepper
- 4 small pieces of streaky bacon
- 6 tablespoons (90 ml) cooking oil
- 7 oz (200 g) mirepoix (chopped onion, carrot, and celery)
- 6 tablespoons (90 ml) Marc de Bourgogne (a type of aged pomace brandy from Burgundy)
- 3/4 cup (175 ml) tannic red wine
- 1 cup (250 ml) black currant juice
- 3/4 cup (175 ml) unthickened brown game-meat stock, or store-bought demi-glace
- 7 oz (200 g) black currants
- 1/3 cup (80 g) butter

Sautéed Partridge with Black Currant Sauce

4 servings • Cooking time: 30 to 40 minutes • 3

There is no comparison between farmed and wild partridge. Wild partridge is better even though farmed partridge is better fed, but wild partridge needs to be marinated 24 to 36 hours (see Basic Recipes for the marinade, p. 136). It is important to choose your equipment wisely for cooking partridge. Since it needs high heat for the entire cooking time, a sauté pan is ideal. You can eat every part of this fine game bird. Even better, you're allowed to eat it with your hands.

Preparation: 20 Minutes

- Season the inside and outside of the partridges. Wrap them with the bacon to keep the breasts from drying out (the fat will baste the meat during cooking) and tie to secure. Heat the oil in the sauté pan and place the partridges in the pan on their backs. Roast in the oven for 8 to 10 minutes at 490 °F (255 °C). Baste with a spoon occasionally. Take the partridges out of the sauté pan, remove any twine from wrapping the bacon, but leave the fat in the bottom of the pan. Add the mirepoix and put the partridges back into the pan. Put everything back into the oven at 300 °F (150 °C) to cook until just done. Remove the excess cooking fat and flambé with the brandy.
- Remove the partridges and keep them warm. Add the wine and black currant juice to the pan. Reduce by half and add the game-meat stock. Simmer for 10 minutes. Strain and press well. Finish the sauce with butter. Add the black currants, but make sure the temperature does not exceed 185 °F (85 °C) so the fruit doesn't burst. Adjust the seasoning.
- Leave the partridges on their backs and serve with the sauce on the side along with a finger bowl. Braised cabbage and potatoes are a suitable accompaniment.

Figs

Ficus carica (Figues)
Family: *Moraceae*

The ancient Egyptians knew figs, which were included in their pharmacopoeia. The Greeks believed they were a source of strength and agility.

Fig trees rarely exceed 5 metres in height. The large, fleshy leaves (10 to 20 cm) are very dark green on top and light green underneath. They have 5 to 6 very indented lobes. It is the only plant that produces its fruit inside a closed receptacle. So figs are not actually fruit, in the true sense of the word, but a whole cluster of flowers. Each female flower yields a one-seeded fruit (achene) and it is this fertile part that constitutes the pink, granular flesh inside of figs. We know the Romans used to force-feed their geese with figs. Purple figs are used in cooking and baking, and, generally, white and green figs are dried.

Cooking and Baking

Figs are excellent with fois gras. Baked in the oven, they go well with food that needs sweetening.
Compotes, jams, cakes, syrups, pies, chocolates, and candies.

Therapeutic Uses

Diuretic, demulcent (to soothe and protect), laxative, nutrient, pectoral (chest disorders), tonic.

Duck Fois Gras with Figs and Asparagus

4 servings • Cooking time: 10 minutes • 2

Prepare the figs and asparagus the night before the meal.

Preparation: 20 Minutes

- Reduce the duck stock by half. Arrange the figs side by side in an ovenproof covered container. Pour the reduced, cooled stock over the figs. Cover and simmer at 185 °F (85 °C), until the figs are fork tender. Set aside and keep in the fridge.
- Boil 4 cups (1 litre) of salted water and cook the asparagus until they are crunchy. Shock with ice water to stop them cooking, drain, and keep in the fridge wrapped in a towel.
- The day of the meal, take the figs out in the morning. An hour before serving the fois gras, put the container with the figs into the oven at 325 °F (160 °C).
- Lightly season the slices of fois gras on each side with ordinary salt and pepper. In a non-stick frying pan, sear the fois gras for 30 seconds on each side. Keep them warm on a rack in the oven for just long enough to make the sauce.
- Pour 1 cup (250 ml) of the cooking liquid from the figs into a pan and whisk quickly. Add the asparagus spears and the Armagnac. Adjust the seasoning.
- Put the sauce and the asparagus onto plates. Place the slices of fois gras on top of the asparagus and sprinkle with a bit of sel de Guérande. Place a fig on each plate and serve hot. Slices of brioche go well with this dish.

INGREDIENTS

- 2 1/2 cups (625 ml) unthickened brown duck stock, homemade or store bought
- 4 ripe figs
- 24 green asparagus spears
- 4 slices of duck fois gras, 2/3" (1.5 cm) thick and 5 oz (150 g) each
- Freshly ground white pepper
- Sel de Guérande (sea salt)
- 1/4 cup Armagnac or Cognac

Goat's Milk Rice Pudding with Caramelized Figs and Lemon Thyme

4 servings • Cooking time: 25 minutes • 3

Preparation: 20 Minutes

- In a saucepan, bring the goat's milk and sugar to a boil. Add the rice and lower the heat to a gentle boil. Stir often until the consistency is creamy but the rice is still *al dente*. After cooking, add the lemon thyme. Spread the mixture onto a baking sheet and cover with plastic wrap. Set aside in the fridge.

Garnish

- Cut the figs in half and sprinkle them with sugar. Caramelize them with a kitchen blowtorch, the same as you would for crème brûlée. If you do not have a torch, put the figs under the broiler for a few seconds.
- Use a ring mould to plate some of the goat's milk rice pudding and place the caramelized figs on top.

INGREDIENTS

- 2 1/2 cups (625 ml) goat's milk
- 1/4 cup (60 g) sugar
- 5 1/2 oz (160 g) Arborio rice
- 1 teaspoon (5 g) fresh lemon thyme

Garnish

- 4 figs
- Granulated sugar

INGREDIENTS

- 2 Barbary figs (prickly pear or cactus pear)
- 2 ripe figs
- 1/3 cup (80 g) unsalted butter
- Salt and freshly ground pepper
- 4 rockfish fillets, 4 oz (120 g) each
- 1/4 cup (50 g) flour, sifted
- 5 tablespoons (75 ml) peanut oil
- 1/5 cup (50 g) unsalted butter
- Juice of one lemon
- 1 1/2 oz (45 g) slivered almonds, toasted
- 1/3 cup (80 g) unsalted butter, melted

Rockfish Fillets, Fig Emulsion, and Barbary Figs

4 servings • Cooking time: 8 minutes • 2

Preparation: 20 Minutes

- Cut the Barbary figs in half and extract the juice without crushing the seeds. Set aside. Cut the figs into quarters. Heat the butter in a non-stick frying pan and slowly cook the figs, turning carefully. Season them with salt and pepper. Keep warm.
- Dry the rockfish fillets well. Season with salt and pepper and dredge them in flour. Heat the peanut oil and the butter. Cook the fillets until they are an even golden colour. Turn down the heat and baste occasionally. Stop cooking when the internal temperature reaches 150 °F (65 °C).
- While the fillets are cooking, pour the juice from the Barbary figs and the lemon juice into a blender with the almonds. Blend on high speed and add in the hot butter. Season with salt and pepper.
- Pour the emulsion into the bottom of each plate. Place a rockfish fillet on top and garnish with the figs cooked in butter. You can also add small steamed potatoes.

Goose with Figs

4 servings • Cooking time: 120 to 150 minutes • 3

This recipe needs a preliminary preparation 24 hours before cooking. Ask your butcher to debone the geese. You will need the four thighs for this recipe. The four breasts can be used in another recipe (e.g., Goose Breast with Red Currants, p. 68). Cut the carcass into pieces to make the brown stock (see Basic Recipes for Brown Game-Meat Stock). It is important to add the roux in small quantities, leaving it to boil between additions. If you put in too much at once, you run the risk of turning the sauce into glue.

Preparation: 40 Minutes

- In a heavy-bottom frying pan, heat the peanut oil and brown the thighs, starting skin side down. Put the thighs into a container that is sufficiently large and cover with the wine. Sauté the vegetable mirepoix in the same pan the thighs were browned in. Pour the mirepoix over the thighs. Add the garlic, bouquet garni, juniper berries, cloves, bay leaf, olive oil, figs, salt, and 10 peppercorns. Cover with plastic wrap and leave on the counter for 24 hours or for 48 hours in the fridge.
- Remove the figs from the marinade and set aside. Put the remainder, including the thighs, into a Dutch oven. Add the goose stock and cook in the oven at 195 °F (90 °C) until a knife can be easily inserted and removed from the thighs (about 2 to 3 hours, depending on the size of the thighs). Remove the thighs and keep warm. Strain the cooking liquid and cook the potatoes. Remove and keep warm.
- Using the white roux, thicken the goose jus to your desired consistency. Adjust the seasoning and strain again. Put the goose thighs, potatoes, and figs into the sauce. Simmer for 10 minutes.
- Serve hot in soup bowls. Sprinkle with the croutons and chopped parsley.

INGREDIENTS

- 6 tablespoons (90 ml) peanut oil
- 2 whole geese, 5 1/2 lb (2.5 kg) each
- 4 cups (1 litre) tannic red wine
- 1 Spanish onion, chopped for mirepoix
- 1 carrot, chopped for mirepoix
- 1 celery stalk, chopped for mirepoix
- 4 cloves of garlic
- 1 bouquet garni
- 12 juniper berries
- 1 clove
- 1 bay leaf
- 6 tablespoons (90 ml) olive oil
- 12 oz (360 g) dried figs
- Coarse salt and peppercorns
- 4 cups (1 litre) brown goose stock
- 12 fingerling potatoes
- White roux
- 3 1/2 oz (100 g) small cubed croutons
- 1 oz (30 g) parsley, finely chopped

INGREDIENTS

- 2 1/2 oz (75 g) ground almonds
- 1 cup (240 g) icing sugar
- 1/3 cup (80 g) flour
- 7 egg whites
- 1/2 vanilla bean
- Zest of a lemon
- 7 oz (200 g) brown butter
- 4 figs

Red Wine Reduction

- 1/2 cup (125 ml) red wine
- 1/2 cup (120 g) sugar

Almond Cakes with Figs and Red Wine Reduction

4 servings • Cooking time: 20 minutes

Preparation: 20 Minutes

- Sift the ground almonds, icing sugar, and flour. In a blender on medium speed, blend the ingredients in this order: egg whites, vanilla, zest, and brown butter. Mix just enough to get a uniform mixture. Chill the batter for a few hours (it will keep for 3 days).
- Fill individual greased and floured moulds halfway with the batter. Place a slice of fig on top of each one. Bake at 350 °F (180 °C) for about 12 minutes.

Red Wine Reduction

- In a saucepan on medium heat, reduce the wine and sugar until it becomes syrupy. Serve the almond cakes still warm with the wine reduction drizzled around the outside.

Ground Cherries

Physalis pruinosa (cerises de terre)
Family: *Solanaceae*

Ground cherries are also sometimes called tomatillos, or "amour en cage" (caged love) in French, and they grow in lime or chalky soil. The round calyx holds the fruit until it is ripe. Its genus is *Physalis*, which in Greek is *phusan*, or "inflated." This small sour fruit is being used more and more in Quebec. It's a good idea to remove the somewhat bitter skin. There are three main varieties of the genus *Physalis*, which can cause some confusion. *P. ixocarpa* (tomatillo) yields a green or yellow fruit, prized in Mexican cuisine. The *P. peruviana*, Cape gooseberry, has a fruit the size of a cherry. *P. pruinosa* is a varietal with the smallest fruit.

Cooking and Baking

Ground cherries go well with veal and some poultry. The reduced juice is good with fish such as monkfish and swordfish.

Compotes, jams, cakes, jellies, sorbets.

Therapeutic Uses

Gout, jaundice, lithiasis (stones), swelling, rheumatism, uremia.

Turkey Cutlets Sautéed with Ground Cherry Compote

4 servings • Cooking time: 15 minutes • 2

Preparation: 30 Minutes

- Put the shallots and wine in a sauté pan. Reduce by half. Add the ground cherries, lemon juice, salt, pepper, and cook slowly to make the compote. At the end, add the walnuts. Keep warm.
- Season the turkey cutlets with salt and pepper, then dredge them in flour. Heat 1/2 cup (120 g) of butter in a heavy-bottom frying pan and sear the cutlets, to give them a nice golden colour. Turn down the heat to finish cooking them through. The internal temperature should not exceed 175 °F (80 °C). Take the cutlets out of the pan and keep them warm. Remove excess cooking fat from the pan and deglaze with the port. Add the chicken stock and finish with the remaining butter. Adjust the seasoning.
- Pour the turkey jus into the bottom of each plate and place a cutlet on top. On each cutlet, place a quenelle (dumpling) of the ground cherry compote. Serve with green beans sautéed in butter.

INGREDIENTS

- 2 dried shallots, finely chopped
- 2/3 cup (160 ml) white wine
- 2 cups (300 g) ground cherries
- Juice of one lemon
- Salt and freshly ground pepper
- 1 cup (120 g) walnuts, in pieces
- 4 turkey breast cutlets, somewhat thick, 5 oz (150 g) each
- 1/3 cup (70 g) flour, sifted
- 3/4 cup (180 g) unsalted butter
- 2/3 cup (160 ml) white port
- 6 tablespoons (90 ml) brown chicken stock, homemade or store bought

Veal Sauté with Ground Cherry

4 servings • Cooking time: 50 minutes • 3

Ask your butcher for veal cubes from the shoulder or neck. For a sauté that is not dry, the pieces of meat should have some fat on them so they will be more tender.

INGREDIENTS

- 1 lb 10 oz (800 g) veal, 3/4" (2 cm) cubes
- Salt and freshly ground pepper
- 3 1/2 oz (100 g) cooking oil
- 7 oz (200 g) mirepoix of onion, celery, and carrot, cut into large pieces
- 1 cup (250 ml) dry white wine
- 3/4 cup (175 ml) ground cherry juice
- 1 bouquet garni
- 6 tablespoons (90 ml) tomato paste
- 2 cups (500 ml) unthickened brown veal stock, homemade or store bought
- White roux
- 10 oz (300 g) ground cherries

Preparation: 20 Minutes

- Season the veal with salt and pepper. In a heavy-bottom frying pan, heat the cooking oil and sear the veal to a golden colour. Place the meat in a small Dutch oven, spread the mirepoix over it, and pour in the wine and ground cherry juice. Add the bouquet garni, cover with plastic wrap, and leave on the counter for 24 hours. There is no risk of contamination because of the alcohol in the wine and the acidity of the fruit.
- Heat the Dutch oven on top of the stove. Add the tomato paste and the veal stock. Cook in the oven at 195 °F (90 °C) until the veal is nice and tender.
- Using a skimmer, take out the meat and keep it warm. Strain the cooking liquid. Add the roux a little at a time, to thicken the liquid to the desired consistency. Adjust the seasoning. Add the ground cherries and the cubes of veal. Heat slowly, without bringing to a boil, so the fruit doesn't burst.
- Serve in soup bowls with small steamed parisienne potatoes.

Vegetable Trio with Ground Cherries

4 servings • Cooking time: 30 minutes • 3

Three vegetables, three different ways of cooking them. Parsnips do not need water to cook if they are cut into thin rounds. However, green beans need a lot of salted water to maintain their vibrant green colour. Once they are cooked, they need to be shocked in ice water to remain crisp. Cook the carrots in sticks in a steamer. They should be crunchy, but not raw. Keep the ground cherries at room temperature for an hour or two before cooking so they burst open.

INGREDIENTS

- 1/3 cup (80 g) unsalted butter
- 12 oz (360 g) parsnips, in rounds
- Salt and freshly ground pepper
- 8 oz (240 g) green beans
- 12 oz (360 g) carrots
- 7 oz (200 g) ground cherries
- 3 oz (20 g) chives, minced
- 1/3 oz (10 g) chervil leaves

Preparation: 30 Minutes

- Heat 1 oz (30 g) of butter in a sauté pan. Add the parsnips, 1/4 cup (60 ml) of water, salt and pepper, and cover. Simmer gently so the parsnips steam. They are perfectly done when they are fork tender.
- In a heavy-bottom frying pan, heat the rest of the butter. Add the parsnips, green beans, carrots, and ground cherries. On medium heat, sauté them for 4 to 5 minutes, until the cherries burst open and release their juice. Add the chives.
- Sprinkle the chervil leaves on top and serve hot.

Es siempre este
el antídoto mejor
contra la
yo por

INGREDIENTS

- 3 eggs
- 2 1/2 tablespoons (37.5 g) sugar
- Zest of a lemon
- 2 cups (500 ml) milk
- 7 oz (200 g) flour
- A pinch of salt
- 5 teaspoons (24.5 ml) vegetable oil

Ground Cherry Compote

- 8 oz (240 g) ground cherries
- 1/4 cup (60 g) sugar

Whipped Cream Cheese

- 1/2 cup (100 g) cream cheese
- 1 tablespoon (15 g) sugar
- 1/2 cup (125 ml) heavy cream (35%)

Crêpes, Ground Cherry Compote, and Whipped Cream Cheese

4 servings • Cooking time: 20 minutes

Preparation: 30 Minutes

- Beat the eggs with the sugar and the lemon zest. Add the milk. In a different bowl, sift the flour and the salt. Fold the wet mixture into the dry ingredients. Finish by adding the oil. Refrigerate for 1 hour.

Ground Cherry Compote

- In a saucepan, cook the cherries and the sugar on medium heat to make a thick compote. Refrigerate.

Whipped Cream Cheese

- In a mixer or with a whisk, beat the cheese and sugar. Drizzle the cream in and whisk to soft peaks. Cook the crêpes in a non-stick frying pan. Serve with the ground cherry compote and the whipped cream cheese.

Lychees

Litchi chinensis (Lychee)
Family: *Sapindaceae*

This beautiful tree is from South-East Asia and can reach a height of ten to fifteen metres. The crown is spread out, and hanging branches with evergreen leaves yield clusters of about twenty oval or round fruits. Their red scaly capsule hides a delicate, juicy, fragrant, gelatinous, sweet flesh. Lychees don't ripen after being picked. After the lychee is picked, it must be handled with care.

The name *litchi chinensis* alludes to its Chinese (possibly Cantonese) origins. For centuries, lychees were offered on New Year's Eve as a good luck charm. The taste is considered to be a perfect combination of rose, strawberry, and muscat grape. Langsat (*Lansium domesticum*) is very different from lychee, but can be considered its first cousin. The ripe fruit has a milky juice and the white pulp is translucent and juicy.

Cooking and Baking

Lychees go with many Asian dishes, but their delicate flavour can be enhanced by balancing the sweetness with a bit of acidity.

Lychees can be eaten fresh, in syrups, and in jams.

Therapeutic Uses

Lychee fruit is rich in vitamin C and carbohydrates.

Lychee and Lemongrass Chicken

4 servings • Cooking time: 30 minutes • 2

Preparation: 30 Minutes

- In a Dutch oven, reduce by half the wine, sake, and lemon juice with the lemongrass. Add the lychee juice. Cool.
- Heat the butter in a heavy-bottom frying pan. Season the chicken breasts with salt and pepper. Brown them on both sides. Put them into the lychee stock. Marinate for 4 hours on the counter, so all the flavours are infused into the chicken. Add the chicken bouillon to the marinade. Heat to 194 °F (90 °C), cooking slowly for 20 to 25 minutes. The internal temperature should be 167 °F (75 °C). Remove the chicken breasts at this point and keep them warm. Reduce the cooking liquid by half.
- In a saucepan, reduce the cream by half. Mix the cold water with the thickener in a separate bowl. Add it, a little at a time, to the cooking liquid to achieve the desired consistency (wait for the boil because the starch only begins to work at this point). Add the reduced cream and adjust the seasoning. Add the lychees and chicken breasts and heat for a few minutes to 194 °F (90 °C).
- Serve with rice and sprinkle with chives.

INGREDIENTS

- 6 tablespoons (90 ml) white wine
- 6 tablespoons (90 ml) sake
- Juice of 2 lemons
- 4 stalks of lemongrass
- 1 cup (250 ml) lychee juice, fresh or unsweetened canned
- 1/2 cup (120 g) unsalted butter
- Salt and freshly ground pepper
- 4 skinless chicken breasts
- 2 chicken bouillon cubes, homemade or store bought
- 1 1/2 cups (375 ml) heavy cream (35%)
- 2 1/2 oz (75 g) veloutine or potato starch
- 32 lychees, fresh or unsweetened canned
- 2 oz (40 g) chives, minced

INGREDIENTS

- 1 package baby spinach leaves
- 2 oranges
- 12 to 16 lychees, fresh or unsweetened canned
- 3/4 cup (175 ml) mayonnaise, homemade or store bought
- 2 1/2 oz (75 g) slivered almonds, toasted
- Salt and ground white pepper
- Lemon (optional)
- 1 lb (480 g) whelks, cooked

- 1 cup (250 ml) fresh pineapple juice
- 1 cup (250 ml) fresh lychee juice
- 4 cups (1 litre) light fish fumet (see Basic Recipes)
- 5 tablespoons (75 ml) olive oil
- 10 oz (300 g) mirepoix of onion, carrot, and celery, diced
- 12 small blue crabs
- 3/4 cup (175 ml) white wine
- 1 bouquet garni
- 7 oz (200 g) pommes de terre noisettes (small potato balls)
- 5 oz (150 g) crab meat, diced
- 24 lychees
- 1/3 oz (10 g) parsley, finely chopped
- 3 oz (90 g) small croutons

Whelk Salad with Orange, Lychee and Mayonnaise

4 to 6 servings • 1

Whelks are a large edible mollusk from the Atlantic Ocean.

Preparation: 15 Minutes

- Stem, wash, and spin dry the spinach. Keep in the fridge. Cut the peeled oranges into small pieces. Pit the lychees and cut them the same size as the oranges. Mix the fruit with the mayonnaise and the almonds. Season to taste with salt and pepper. If it isn't tart enough, add the lemon juice. Set aside.
- Cut the cooked whelks and mix with the mayonnaise. Keep in the fridge until it is time to serve them.
- Make a circle of spinach leaves on the plate and pile the whelk salad in the centre. Serve cold.

Crab Soup with Pineapple and Lychees

4 servings • Cooking time: 30 to 40 minutes • 1

Preparation: 60 Minutes

- Mix the fish fumet, pineapple, and lychee juices.
- In a Dutch oven, heat the olive oil and sauté the mirepoix. Add the crabs and wine. Reduce it by half to remove the acidity and pour in the juice mixture. Add the bouquet garni and cook for 20 to 30 minutes. Take the crabs out and chill. Strain and reduce the liquid by 40% while cooking the potatoes in it.
- While the liquid reduces, remove the meat from the crabs and dice. Peel and pit the lychees and cut them into quarters.
- Put the crab meat and the lychees in the bottom of each soup plate. Pour in the broth, with the potatoes on top. Sprinkle with chopped parsley and add the croutons.

Veal Tongue with Lychee Sauce

4 servings • Cooking time: 60 minutes • 2

Preparation: 20 Minutes

- Put the veal tongues into a large bowl and run cold water over them.
- In a Dutch oven, boil 12 to 16 cups (3 to 4 litres) of water with an onion spiked with cloves, a whole carrot, celery stalk, peppercorns, unpeeled garlic, and bouquet garni. Lightly season with salt. Cook for 30 to 40 minutes, enough time for the stock to take on the flavour of the aromatics. Put in the veal tongues and cook at 194 °F (90 °C), until they are fork tender.
- While they are cooking, make the sauce. Mix the mustard, egg yolks, and red wine vinegar. Season with salt and pepper. Slowly drizzle in the sunflower oil to make a mayonnaise. Add the parsley, lychees, and the hardboiled egg white. Adjust the seasoning.
- Remove the skin from the tongues, cut them into thick slices, and place them on a plate. Pour the lychee sauce on top. Small pommes de terre noisettes are a nice side dish.

Marinated Lychees with Vanilla Lime Granita

4 servings

Preparation: 10 Minutes

- Boil the water, sugar, and vanilla. Chill the syrup and add the lime juice. Pour into a container and keep in the freezer.
- Boil the water, sugar, and lime juice. Pour over the peeled, pitted, and halved lychees. Marinate for a few hours in the fridge.
- Put the lychees in a small bowl or dessert cup and garnish with the granita. Take the granita out of the freezer 2 minutes before plating.

INGREDIENTS

- 2 veal tongues
- 1 onion
- 1 clove
- 1 carrot
- 1 celery stalk
- 10 black peppercorns
- 4 cloves of garlic, unpeeled
- 1 bouquet garni
- Salt

Lychee Sauce

- 1/2 oz (15 g) Dijon mustard
- 2 egg yolks
- 3 tablespoons (45 ml) red wine vinegar
- Salt and ground white pepper
- 3/4 cup (175 ml) sunflower oil
- 3 1/2 oz (100 g) parsley, chopped
- 3 1/2 oz (100 g) lychees, chopped
- 1 egg white, hardboiled and chopped

Granita

- 3/4 cup (175 ml) water
- 7 oz (200 g) sugar
- 1/2 vanilla bean, split and the seeds scraped out
- 1 cup (250 ml) lime juice

Lychees

- 2 cups (500 ml) water
- 7 oz (200 g) sugar
- Juice of 2 limes
- 24 lychees

MIRABELLE PLUM
WILD PLUM
DAMSON PLUM
GREENGAGE PLUM

Plums

Prunus domestica (Prune)
Family: *Rosaceae*

There are hundreds of plum trees, but the origins of this tree have been lost over time. Its complex evolution has produced fruits of many colours and shapes. Plums can be yellow, green, red, purple, or black, oblong, oval, or spherical. The uses of the fruit depend on the flesh, which can be either soft or firm.

Damson plums are purple or yellow. The flesh is rich, slightly tart, and very pleasant. Mirabelles are small, spherical, and golden-yellow dotted with red. The fruit is firm, sweet, and fragrant, with a unique taste. Avalon plums are large and oblong. They are dark purple with very firm flesh and an intense, sweet flavour. The spherical greengage plums are green with tender, fragrant flesh. All of these plums are found in the supermarket.

Prunes are dried plums. Perdrigon plums from Spain, for example, become like coins when sun dried and peeled with a knife. Prunes from Agen and Tours are highly esteemed, but those from California and Australia are serious contenders. Prunes are meant to be cooked for a long time with red wine and can be paired with white meat or a fatty fish.

Other Examples of *Prunus* in Quebec
Sand cherry, *Prunus depressa*
Canada plum, *Prunus nigra*
Pin cherry, *Prunus pensylvanica*
Black cherry, *Prunus serotina*
Choke cherry, *Prunus virginiana*

Other Varieties of Plums in Quebec
Blue damas, purple damas, local plum, île d'Orléans plum, Agen (sugar) plum, Lombard plum, Mount-Royal plum, greengage plum, Mirabelle plum.

Cooking and Baking
Cookies, jams, compotes, cakes, syrups, pies.

Therapeutic Uses
Anaemia, anxiety, atherosclerosis, constipation, gout, rheumatism, stress, hepatic disorders.

Duck Livers with Wild Plums

4 servings • Cooking time: 12 minutes • 2

Wild plums are always more tart. If the plums you have are too sweet, you'll need to add lemon juice.

INGREDIENTS

- 1 kg (480 g) fresh duck livers
- 7 oz (200 g) wild plums or small sour plums
- 5 tablespoons (75 ml) olive oil
- 3 dried shallots, finely chopped
- 2 cloves of garlic, chopped
- Salt and freshly ground pepper
- 10 oz (300 g) bean sprouts
- A pinch of grated nutmeg
- 1/2 cup (120 g) unsalted butter
- Salt and freshly ground pepper
- 1/3 cup (80 ml) Cognac
- 2/3 cup (160 ml) white wine
- 3/4 cup (175 ml) thickened brown chicken stock, homemade or store bought
- 3 oz (90 g) parsley, chopped

Preparation: 20 Minutes

- Clean the duck livers. Cut them into 1/2" (1 cm) cubes. Place them on paper towels and set aside in the fridge.
- Pit the plums and cut them into 1/2" (1 cm) strips. Heat the olive oil in a sauté pan. Sweat the shallots and garlic. Add the plums and season with salt and pepper. Cook gently, covered, at 175 °F (80 °C) to keep the plums firm. Set aside in the pan.
- In a heavy-bottom frying pan or a wok, heat the oil. Sear the bean sprouts and sauté them quickly. Season with salt and pepper. Add the nutmeg. Keep the bean sprouts warm.
- Season the duck livers with salt and pepper. In a heavy-bottom frying pan, heat the unsalted butter and sear the livers. Keep them rare, otherwise they will be tough. Flambé with cognac and remove the livers from the pan. Keep warm. Add the wine and reduce by half. Add the chicken stock. Adjust the seasoning and quickly add in the duck livers, bean sprouts, and plums. Sprinkle with chopped parsley and serve in soup bowls.

INGREDIENTS

- 2 cups (500 ml) plum juice
- Salt and pepper
- 1/4 cup (60 ml) white vinegar
- 4 calf brains, soaked
- Thick cream (35%)
- 3/4 cup (175 ml) sour cream
- 1/4 cup (60 ml) vodka
- Chervil leaves

Veal Brain Mousse with Sour Cream and Plum Juice

4 servings • Cooking time: 20 minutes • 3

Preparation: 40 Minutes

- Heat the plum juice with the white vinegar to 175 °F (80 °C). Season it very lightly with salt and pepper. Immerse the brains in it. Simmer for 40 to 45 minutes to reach an internal temperature of 154 °F (68°C). Let the brains cool in the cooking liquid.
- The next day, reduce the cooking liquid by 90% and cool (if in a hurry, it can be chilled in the freezer). Dry the brains well and process in a blender or food processor. Weigh and put in the fridge.
- Measure thick cream (35%) to half the weight of the brains and whip it. Gently fold together the whipped cream and the brains. Adjust the seasoning and put the mixture in the freezer until it is very cold.
- Mix the sour cream with the vodka and the reduced plum juice. Season with salt and pepper.
- Put a spoonful of the sour cream mixture in the centre of the plate. Using two warm soup spoons, form quenelles (dumplings) with the brain mousse and place them on top of the sour cream. Sprinkle with chervil leaves.

Plum and Mascarpone Shortcakes

4 to 6 servings • Cooking time: 10 minutes

Preparation: 30 Minutes

- Mix the dry ingredients with the butter until crumbly. Add the cream and mix until the dough comes together in a ball. The less the dough is handled, the better the texture will be. Roll out the dough on a lightly floured surface. When it has been rolled to a thickness of 3/4" (2 cm), cut biscuits with a round cookie cutter about 2 1/2" (6 cm) in size. Put the shortcake biscuits into the fridge for at least 1 hour.
- Lightly brush the biscuits with cream and sprinkle with a thin layer of sugar. Bake at 350 °F (180 °C) for about 10 minutes.

Mascarpone Cream

- In a mixer or with a whisk, mix the cheese and icing sugar on medium speed. Gradually add the cream and whisk to soft peaks.

Plum Jam

- In a saucepan, cook the plums with 2 tablespoons of sugar and the lemon juice for about 5 minutes. Take out the plums and boil the syrup. Add the pectin, mix in the teaspoon of sugar, and boil again for 1 minute. Put the plums back in the pot and continue to cook for 1 minute. Chill.
- Cut the shortcake biscuits in half and spread with the plum jam. Add a bit of the mascarpone cream and top with the other half of the shortcake biscuit.

INGREDIENTS

Biscuits

- 1 1/3 cups (240 g) flour
- 2 tablespoons (30 g) sugar
- 1 teaspoon (5 g) baking powder
- A pinch of salt
- 1/3 cup (80 g) butter
- 5 oz (150 g) heavy cream (35%)

Mascarpone Cream

- 4 oz (120 g) mascarpone cheese
- 1/2 cup (125 ml) heavy cream (35%)
- 3 tablespoons (45 g) icing sugar

Plum Jam

- 7 oz (200 g) plums, diced
- 2 tablespoons (30 g) sugar
- Juice of one lemon
- 1 teaspoon (5 ml) pectin
- 1 teaspoon (5 g) sugar

Raspberries

Rubus idaeus (Framboises)
Family: *Rosaceae*

The raspberry plant is a deciduous shrub. Each year it sprouts new shoots that live for two years. The second year, they bear fruit, dry up, and die. The word "raspberry" comes from the Frankish (the language spoken by the Germanic Franks) word *branbasia*—berry—which referred to the fruit of the shrub. Cultivating raspberry bushes goes back to the Middle Ages, but our prehistoric ancestors also appreciated this delicious fruit. Raspberries, like strawberries, contain salicylic acid, a valuable compound that fights rheumatism and gout. Raspberries retain nearly all their benefits, even when frozen.

In Quebec, we are familiar with the red raspberry, black raspberry, flowering raspberry, and hairy raspberry. Black raspberries can be either cultivated or wild. The flavour of the fruit is more intense and yet quite delicate. Flowering raspberries need a lot of patience to pick as the fruit is scattered throughout the plant. Its Iroquois name means "old shoe," because the *coureur de bois* used to put the leaves in their leather shoes to protect their feet. The fruit of the hairy raspberry falls between red raspberries and blackberries.

Other Varieties

Dwarf raspberry, *Rubus pubescens* (Catherinette)
Black raspberry, *Rubus occidentalis* (Framboisier noir)
Flowering raspberry, Rubus odoratus (Ronce odorante)

Cooking and Baking

Some sauces or a jus for calf's liver, eggs, tongue, and brain.
Various neutral-tasting fish.
Compotes, jams, jellies, vinegar.
Eau-de-vie and liqueur.
Chocolates, cakes, fruit paste, and sorbets.

Therapeutic Uses

Anhidrosis (significant lack of perspiration), anxiety, asthenia, constipation, dermatosis, dyspepsia, gastro-intestinal difficulties, fever, gout, rheumatism.

Calf's Liver with Raspberry Vinegar

4 servings • Cooking time: 5 minutes • 3

You can adapt this recipe and use whatever kind of liver you like. For the best results, the slices of liver should all be the same size. Rare liver is not appetizing, and when it is well done, it is flavourless. Only liver cooked "à pointe," that is, still pink, has the best flavour. Use a cast-iron or heavy-bottom frying pan to sear the liver.

Preparation: 20 Minutes

- Reduce the raspberry vinegar by 75%. Add the veal stock and simmer for 4 to 5 minutes. Season with salt and pepper and set aside.
- Dust a cookie sheet with flour. Heat 2/3 cup (160 g) of butter in a frying pan. Flour the slices of liver quickly and shake off the excess. Put them in the hot butter and season with salt and pepper. Cook for 2 to 3 seconds on each side, depending on the thickness. The liver should be "à pointe" or just done. Plate the slices of liver.
- Remove the excess butter from the pan and deglaze with the raspberry vinegar. Add the reduction and finish with 1/3 cup (80 g) of butter. Add the chives. Spoon the sauce around the slices of liver. Serve with mashed potatoes.

INGREDIENTS

- 2/3 cup (160 ml) raspberry vinegar
- 3/4 cup (175 ml) unthickened veal stock, homemade or store bought
- Salt and freshly ground pepper
- 2/3 cup (160 g) flour
- 1 cup (240 g) unsalted butter
- 4 slices veal liver, 5 oz (150 g) each
- 2 dried shallots, finely chopped
- 5 tablespoons (75 ml) raspberry vinegar
- 2 oz (60 g) chives, minced

Veal Brain Flan, with Veal and Raspberry Jus

4 servings • Cooking time: 30 minutes • 3

A lot of people don't like eating brains, but why? I think perhaps the appearance repels them. But veal brain is excellent for your health and the flavour is quite delicate.

Preparation: 35 Minutes

- Soak the brains by running a thin stream of cold water over them for a couple of hours to remove the blood. Remove all the membrane that surrounds them while they are under the water. When all impurities are removed, dry the brains off well with paper towels. Put them into a blender. Add the salt, pepper, egg white, and cream. Blend for 1 minute and strain. Leave in the fridge for 1 hour to rest.
- With a brush, grease 4 ramekins and put them into the fridge to harden the butter. Pour the mixture into the ramekins. Cook them for about 30 minutes in a bain-marie at 350 °F (180 °C), just until the flan becomes firm.
- While they're baking, finely chop the onion and put it in a saucepan. Pour in the raspberry pulp and vinegar. Reduce by half and add the veal stock. Cook for 10 minutes and strain. Add the cream. Adjust the seasoning and set aside.
- Pour a good amount of the sauce in the bottom of each plate. Place the flan in the middle. Arrange the raspberries and the mint leaves around the plate and on top of the flan.

INGREDIENTS

- 4 small calf's brains (1 1/4 lb / 600 g in all)
- Salt and ground white pepper
- 1 egg white
- 1 cup (250 ml) heavy cream (35%), very cold
- 1/4 cup (60 g) unsalted butter, softened
- 2 oz (60 g) small Spanish onions
- 2/3 cup (160 ml) raspberry pulp
- 5 tablespoons (75 ml) raspberry vinegar
- 3/4 cup (175 ml) unthickened veal stock, homemade or store bought
- 1/2 cup (125 ml) heavy cream (35%)
- 16 fresh raspberries
- 12 fresh mint leaves

INGREDIENTS

- 13 oz (400 g) very ripe raspberries
- 1 bottle of white Bordeaux wine (Graves)
- 3 3/4 cups (900 g) white granulated sugar

Raspberry Wine

Preparation: 15 Minutes

- Put the raspberries and the wine into a large-mouth Mason jar. Close tightly and macerate for 24 hours in the fridge. Strain, pressing the raspberries through the sieve. Add the sugar and bring to a boil, stirring frequently.
- Cool. Strain again and pour into opaque bottles. Keep at cellar temperature for 2 months. Drink this wine chilled.

Raspberry Vinegar

Preparation: 15 Minutes

- When you make jelly from raspberries or other berries with seeds, keep the seeds in an opaque jar. Fill the jar with raspberry vinegar or good quality red wine vinegar. Close the jar and keep at room temperature for 6 months.

INGREDIENTS

Crumble

- 2/3 cup (120 g) flour
- 1/2 cup (120 g) cold butter, cubed
- 1/2 cup (120 g) sugar
- 3 1/2 oz (100 g) ground almonds
- Zest of a lemon
- 1/2 teaspoon freshly ground pepper

Raspberry Jam

- 10 oz (300 g) raspberries
- 1/3 cup (80 g) sugar
- 5 g (1 tsp) pectin
- 5 g (1 tsp) sugar

Yogurt

- 3 cups (750 g) plain yogurt
- 1/4 cup (60 g) sugar

Raspberry Jam, Yogurt, Pepper, and Lemon Crumble

6 to 8 servings • Cooking time: 10 minutes

Preparation: 20 Minutes

- For the crumble, blend all the ingredients until the dough comes together in pea-sized pieces. Spread it out on a baking sheet and cover with parchment paper. Refrigerate for 1 hour. Bake at 350 °F (180 °C) for about 6 minutes, turning the crumble over a few times to bake evenly. Let rest before using.

Raspberry Jam

- Cook the raspberries and sugar on medium heat until the mixture boils and the berries release their juice.
- Strain. Bring the juice to a boil again. Whisk in the rest of the sugar and the pectin. Cook for 1 minute. Add the raspberries and cook again for 1 minute more. Set aside in the fridge.

Yogurt

- Line a sieve with a coffee filter and place it in a bowl. Pour in the yogurt and drain overnight in the fridge. The next morning, add the sugar to the yogurt.

Assembly

- In a parfait glass, alternate layers between the jam, yogurt, and crumble.

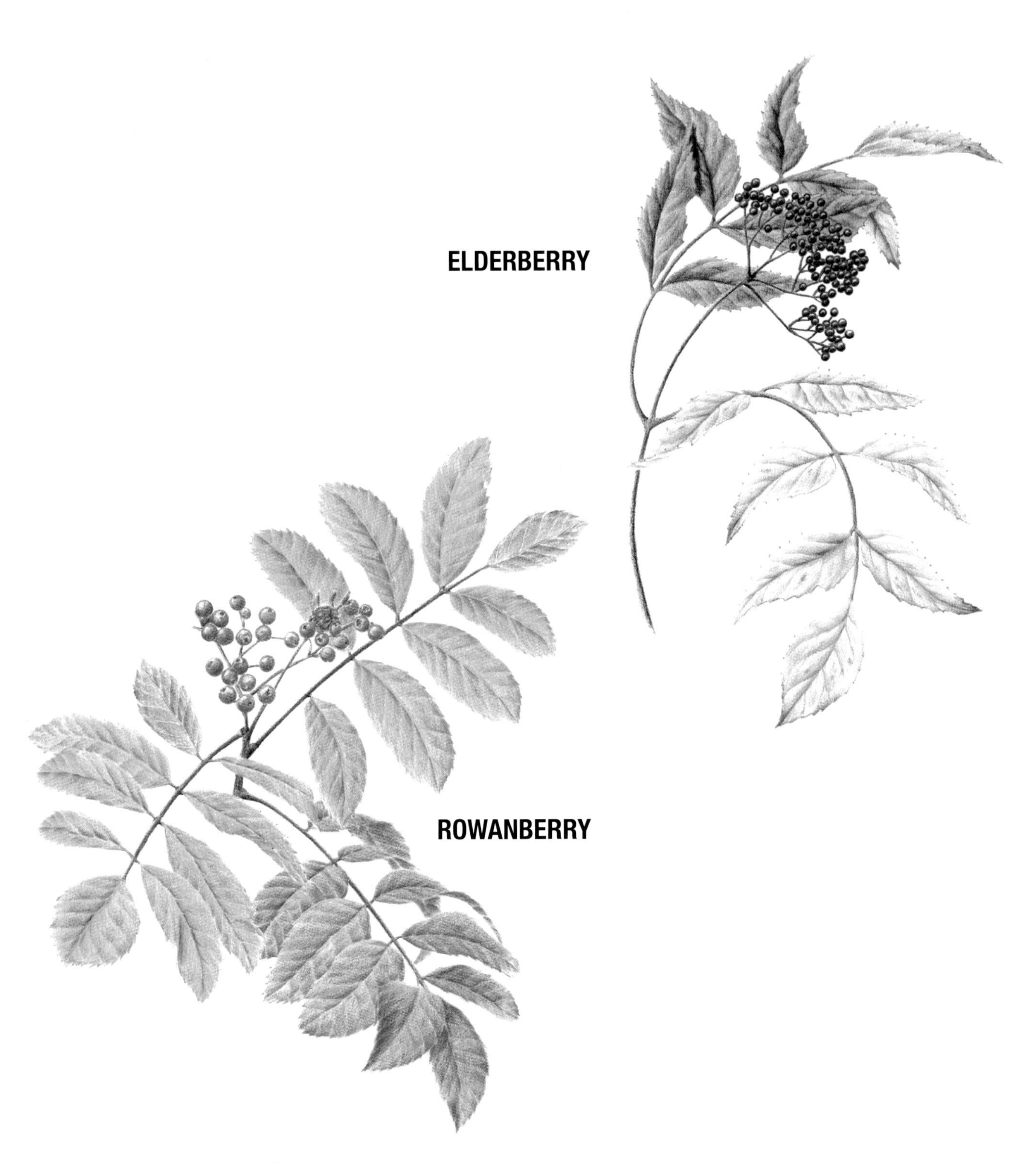
ELDERBERRY
ROWANBERRY

Rowanberries | Elderberries

Family: *Rosaceae*
Family: *Caprifoliaceae*

The rowan is a small tree with bronze bark and white flowers that smell like hawthorn. During the summer, berries gradually replace the flowers. They resemble tiny apples, but are inedible when raw. Only birds can digest them. The American mountain-ash is a tree that is not cultivated a lot because it grows spontaneously on rock faces and escarpments. Its flowers are small and white, arranged in bushy bouquets. In October, they transform into clusters of small orange fruit. They are corms, or sorbs (from *sorbus*), which have a bitter taste if they are not left to ripen on straw until almost rotten, like medlars (*Mespilus germanica*), for example, which are similar to rosehips.

The Latin name for the elder, *sambucus*, comes from the Greek *sambuke*, a triangular harp, but in Latin it is a flute carved from the wood of the elder tree. Some traces of elder can be found on sites that date back to the Stone Age. Do not confuse black elderflower berries with the toxic dwarf elderflower (*Sambucus ebulus*), which is a tall, herbaceous plant, whereas black elder is a tree. Today we are rediscovering through phytotherapy that elder is a "healthy" tree. All the parts can be used, including the bark. The sticky dried flowers, with antispasmodic properties, make a pleasant herbal tea.

Other Rowan Trees

American mountain-ash, *Sorbus Americana* (Sorbier d'Amérique)
Rowan or Mountain-ash, *Sorbus aucuparia* (Sorbier des oiseaux)
Showy mountain-ash, *Sorbus decora* (Sorbier plaisant)

Other Elder Trees

American elderberry, *Sambucus canadensis* (Sureau blanc)
Black elder, *Sambucus nigra* (Sureau noir)
Red elderberry, *Sambucus pubens* (Sureau rouge)

Cooking and Baking

ROWANBERRIES: Jellies, marmalades, eau-de-vie, and liqueur
ELDERBERRIES: This fruit can accompany all meats and poultries.
It is used to make eau-de-vie and alcohol for deglazing.
Compotes, jams, cakes, jellies, ice creams, sorbets.

Therapeutic Uses

ROWANBERRIES: Antiscorbutic, astringent, diuretic.
ELDERBERRY: Berries: Antirheumatism and lapactic (purgative).
Flowers: Depurative (purgative), diuretic, galactogine (increases breast milk), sudorific products.

Roasted Pork Tenderloin, Gentian Wine Sauce

4 servings • Cooking time: 20 minutes • 3

In Quebec, we don't grow anything from the gentian family, but we do have "Suze," an aperitif liqueur that is bitter like rowanberries.

Preparation: 30 Minutes

- Cut the tips from the tenderloins and set aside for a recipe requiring thin slices of pork. Season the tenderloins with salt and pepper. In a sauté pan, heat the oil and sear the tenderloins to brown them.
- Roast in the oven for 2 to 3 minutes at 400 °F (200 °C). Lower the heat to 300 °F (150 °C). Baste frequently using a spoon. Roast to an internal temperature of 167 °F (75 °C). Take the tenderloins out of the oven and set aside.
- Remove the excess cooking fat from the pan. Add the shallots and flambé with the Suze until the alcohol evaporates. Add the veal stock and simmer for 1 to 2 minutes. Finish with butter and adjust the seasoning.
- Pour the sauce on the bottom of the plates. Cut grenadins (small scallops) of the pork and place on top of the sauce. Mixed vegetables go well with this dish.

INGREDIENTS

- 2 pork tenderloins, 10 to 13 oz (300 to 400 g) each
- Salt and freshly ground pepper
- 1/4 cup (60 ml) cooking oil
- 1 cup (250 ml) Suze (a gentian aperitif)
- 1 cup (250 ml) thickened brown veal stock
- 4 dried shallots, finely chopped
- 3 1/2 oz (100 g) unsalted butter
- Salt and freshly ground pepper

Roasted Venison, Shallot Compote, and Rowanberry Jus

4 servings • Cooking time: 50 minutes

In Quebec, there are many different sorbs, but two are most common: the snowy mountain-ash and the rowan tree. The berries are bitter and should be picked after the first frost. They will keep in the freezer until needed. When cooking them, combine them with a very sweet fruit or make a gastrique. This recipe might seem complicated but it is extremely simple.

Preparation: 30 Minutes

- Season the venison well with salt and pepper. In a roasting pan, heat the oil and uniformly sear the roast. Cook in the oven for 5 to 6 minutes at 400 °F (200 °C). Reduce the temperature to 350 °F (180 °C). Baste frequently. Halfway through cooking (118 °F/48°C), as indicated by a thermometer, put the mirepoix vegetables, peppercorns, juniper berries, garlic, bay leaf, thyme, and the venison bones around the roast.
- Continue cooking. When the internal temperature of the roast reaches 130 °F (54 °C), take it out of the oven and keep it warm. Remove the excess fat from the pan. Pour in the wine and water and reduce by 80%. Strain and add the game-meat stock. Slowly add the rowanberry gastrique, tasting often to get a balance between the bitter, sweet, and sour flavours.
- Finish the sauce with butter. Adjust the seasoning.
- Pour the sauce on the bottom of the plates. Place a thin slice of the roast on top. Serve with mashed potatoes and seasonal baby vegetables.

INGREDIENTS

- 2 lb 10 oz (1.2 kg) venison roast (shoulder or thigh; outside round or knuckle)
- Salt and freshly ground pepper
- 2/3 cup (160 ml) cooking oil
- 1 carrot, in small cubes for mirepoix
- 1 onion, in small cubes for mirepoix
- 1 celery stalk, in small cubes for mirepoix
- 8 black peppercorns
- 8 juniper berries
- 1 clove of garlic
- 1 bay leaf
- 1 sprig of fresh thyme
- Venison bones, in small pieces
- 3/4 cup (175 ml) red wine
- 1 cup (250 ml) water
- 1 cup (250 ml) brown game-meat stock (see Basic Recipes), homemade or store bought
- 1/2 cup (120 g) unsalted butter

Gastrique

- 1 cup (150 g) rowanberries
- 7 oz (200 g) granulated white sugar
- 2/3 cup (160 ml) red wine vinegar

INGREDIENTS

- 30 cups (7.5 litres) white wine
- 13 oz (400 g) sugar cubes, crushed
- 3 cups (300 g) white elderflowers, fresh or dried

Vin (Muscat) de Lunel

For this recipe, fresh or dried white elderflowers are used. They impart a pleasant smokiness to the wine.

- Pour the wine into 1 or 2 large-mouth Mason jars. Add the crushed sugar and the elderflowers. Close with the airtight lid.
- Infuse for 8 to 10 days. Strain and then strain again through a coffee filter. Pour into opaque bottles and cellar.

SQUASHBERRY

Squashberries | Highbush Cranberries

Viburnum edule (Eastern and Northern Quebec) (Pimbina)
Viburnum trilobum (Western and Central Quebec) (Viorne comestible)
Family: *Caprifoliaceae*

Cooking and Baking
Flans and compotes for game meat; jams.

The Latin name, *viburnum*, means "limber," an allusion to the flexibility of the branches of the highbush cranberry and squashberry.

The taste of this fruit resembles cranberries. It releases an unpleasant odour that is nothing like the taste. The berries are used to make wine, jelly, jams, syrups, and pies. It's best to pick the fruit after the first frost or in winter.

The hedges can be one to four metres tall and are loaded with pretty little red or black berries that people were quick to pick. According to Frère Marie-Victorin, they are a favourite of robins needing sugar for the long flight south in the autumn. A Native tradition suggests pouring squashberry syrup onto a bed of clean snow and eating it the same as maple syrup.

Coconut Beef with Spicy Paste and Squashberry Jus

4 servings • Cooking time: 20 minutes • 2

Preparation: 20 Minutes

- Toast the grated coconut on low heat in a wok, with no oil, stirring constantly. Grind in a blender.
- Add the rest of the ingredients for the spicy paste except the sunflower oil. Blend to a purée.
- Heat the sunflower oil in the wok and sauté the spicy paste, until it releases a pleasant fragrance. Add the coconut, pumpkin, beef, and the squashberry juice. Simmer covered until the beef is tender. If needed, add water or light beef stock while cooking.

INGREDIENTS

- 8 oz (240 g) coconut, freshly grated
- 3 cups (750 ml) coconut milk
- 3 1/2 oz (100 g) pumpkin, cubed
- 1 1/2 lb (1.2 kg) beef, filet mignon, in 3/4” (2 cm) cubes
- 6 tablespoons (90 ml) squashberry juice

Spicy Paste

- A pinch of freshly ground white pepper
- A 4” (10 cm) piece of fresh turmeric, peeled and minced
- A 4” (10 cm) piece of fresh ginger, peeled and minced
- 1 teaspoon salt
- 1/4 teaspoon (1 g) coriander seeds
- 1/8 teaspoon (1/2 g) ground cumin
- 1/2 cup (125 ml) sunflower oil

INGREDIENTS

- 4 veal chops, bone in, 6 oz (180 g) each
- Salt and freshly ground pepper
- 5 tablespoons (75 ml) cooking oil
- 2 1/2 oz (75 g) unsalted butter
- 3/4 cup (175 ml) unthickened brown veal stock
- 5 oz (150 g) squashberry jelly
- 1/2 cup (120 g) unsalted butter

Veal Chops with Squashberry Jelly

4 servings • Cooking time: 12 minutes • 2

Squashberries should be harvested after the first frost of autumn, when they are the juiciest. This recipe includes instructions for making squashberry jelly that is added to the sauce or can be served on the side.

Preparation: 10 Minutes

- Season the veal chops with salt and pepper. In a cast-iron or heavy-bottom frying pan, heat the oil and butter. Sear the veal chops on both sides, lower the heat and cook, basting the meat regularly. Remove the cooking fat. Pour in the veal stock and simmer for 2 to 3 minutes with the veal chops. Remove the chops and keep them warm. Whisk in the squashberry jelly and cook for a few minutes. Finish the sauce with butter and adjust the seasoning. Serve with a rutabaga purée.

Squashberry Jelly

- Extract the juice from the squashberries with a juicer or put the berries in a pot to heat and crush them. When they have burst, strain them. Leave to drain well. Measure the quantity of juice and add the granulated sugar at a ratio of 75% sugar to juice (example: 750 grams sugar to 1 litre juice). Cook the whole thing, skimming frequently, until a candy thermometer reads 219 °F (104 °C). While still hot, pour into jars and seal. Turn them until they cool completely. Keep them in a cold cellar or in the freezer.

Dried Fruit and Squashberry Wine

Tartaric acid, tannin, and brewer's yeast can be obtained in speciality wine-making stores.

INGREDIENTS

- 3 1/3 lb (1.5 kg) squashberries, frozen or dried
- 3 1/3 lb (1.5 kg) dried raisins
- 3 1/3 lb (1.5 kg) dried apricots
- 3 1/3 lb (1.5 kg) dried prunes
- 3 1/3 lb (1.5 kg) dried figs
- 2 lb (1 kg) dried apples
- 5 1/2 cups (1.3 kg) granulated white sugar
- 5 tsp (25 g) tartaric acid
- 1/2 tsp (3 g) tannin
- 3 tsp (16 g) brewer's yeast
- 25 pints (28 litres) distilled water

Preparation: 15 Minutes

- Put all the fruit in a large plastic pail. Boil 28 cups (7 litres) of water and pour it over the fruit. Mix it all well and macerate for 5 hours. Strain the juice and pour into a large jug. Pour 28 cups (7 litres) of boiling water onto the fruit and leave for 3 hours. Strain the juice again and add to the jug. Repeat these steps again two more times.
- Melt the sugar in a bit of water; add the tartaric acid, tannin, and the brewer's yeast. Pour into the jug and stir. Leave to ferment. Carefully strain and put into opaque bottles. Cellar it.
- Keep the fruit in small freezer bags. It can be used for another recipe.

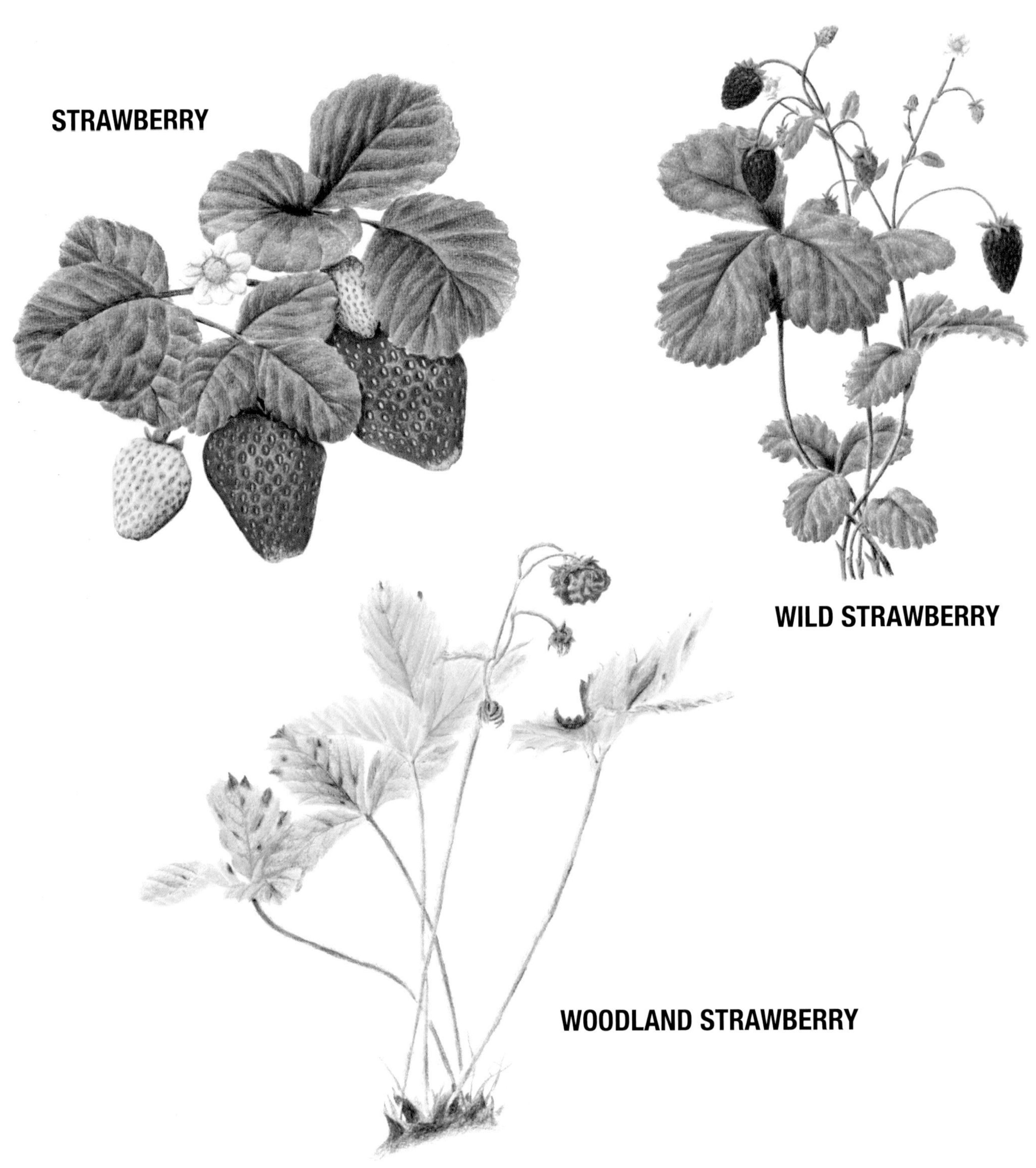
STRAWBERRY
WILD STRAWBERRY
WOODLAND STRAWBERRY

Strawberries

Fragaria vesca (Fraise)
Family: *Rosaceae*

Up until the 17th century, only wild strawberries were known. Since they were so small, only children picked them. The larger varieties (*Fragaria chiloensis*) were brought from Chile to France by Amédée-François Frézier. In Quebec, there are two varieties of strawberries (*F. americana* and *F. virginiana*) that yield excellent fruit. Strawberries are not really a fruit in the botanical sense. The fruit is actually the "achene," the small seeds on the surface of the strawberry. The delicious flesh is an enlargement of the floral peduncle, attributable to hormones secreted by the achenes. The famous naturalist Linnaeus had a case of gout that was healed by strawberry leaves.

Cooking and Baking

Accompanies some vegetables and fish, and sweet and sour sauce.

Chocolate, cakes, ice creams, granitas, mousse, petit fours, sorbets, pies, and tarts.

Therapeutic Uses

Internal use: bactericidal, detoxification, diuretic, gout, hypotensor, laxative, refresher, tonic

External use: astringent, revitalizer.

Cornish Hens Stuffed with Chanterelles with a Strawberry and Red Currant Jus

4 servings • Cooking time: 60 minutes • 4

It is not easy to use strawberries in cooking because of their high sugar content. That's why red currants are included in this recipe. They give it a slight tartness that will be welcome. If you are not used to deboning poultry, ask your butcher to dress the small Cornish hens for you.

Preparation: 60 Minutes

- Wash well, drain, and dry off the currants on paper towels. In a frying pan, heat the olive oil and sear the mushrooms. Add the shallots. Season with salt and pepper. Cook until all the liquid evaporates. Add the Cognac and cool. Add the currants to the poultry stuffing.
- Spread open the hens. Season the interior with salt and pepper. Make a ball of stuffing and reshape the hens around it. Wrap them in aluminum foil.
- Chose a baking dish where the hens can be close together. In this dish, heat the olive oil and put in the foil-wrapped hens. Season the exterior with salt and pepper and roast for 10 minutes at 490 °F (255°C). Take the hens out of the oven and remove the aluminum foil. Baste with the cooking liquid and put them back in the oven at 350 °F (180 °C) until the internal temperature reaches 167 °F (75 °C).
- While the hens are roasting, extract the juice from the strawberries and the red currants. Strain. Heat and reduce by half. Add the poultry stock and the butter. Adjust the seasoning. If it is too sweet, add lemon juice.
- Place the hens in the centre of the plates and sauce. Serve with pomme noisettes (small balls of sautéed potatoes).

INGREDIENTS

- 7 oz (200 g) fresh Chanterelles
- 5 tablespoons (75 ml) olive oil
- 2 dried shallots, finely chopped
- Salt and freshly ground pepper
- 1/4 cup (60 ml) Cognac
- 7 oz (200 g) poultry stuffing
- 4 small Cornish hens or other chicken, deboned through the back
- 6 tablespoons (90 ml) olive oil
- 7 oz (200 g) strawberries
- 7 oz (200 g) red currants
- 2/3 cup (160 ml) unthickened brown poultry stock, homemade or store bought.
- 1/2 cup (120 g) unsalted butter

INGREDIENTS

- 3 1/2 oz (100 g) duck fat
- 8 oz (240 g) diced pork belly
- 10 oz (300 g) chicken livers, in pieces
- 2 1/2 oz (75 g) fresh mushrooms, minced
- 1 oz (30 g) shallots, chopped
- 3 1/2 oz (100 g) fois gras, diced (optional)
- 6 tablespoons (90 ml) Cognac
- 4 egg yolks
- 3/4 teaspoon (4 g) salt
- A pinch of pepper
- A pinch of four-spice (a mixture of ground pepper, ginger, nutmeg, and cloves)

Preparation: 15 Minutes

- 2 slices of bread, in small pieces
- 2/3 cup (160 ml) heavy cream (35%)
- 5 oz (150 g) pork shoulder
- 5 oz (150 g) veal
- 5 oz (150 g) chicken livers
- 1 egg
- Salt and ground white pepper

Stuffing au Gratin

6 to 8 servings • Cooking time: 30 minutes

This stuffing is ideal for special occasions.

Preparation: 30 Minutes

- In a pan, heat the duck fat and brown the pork belly. Remove the pork and add the chicken livers. Brown them. Put the browned pork belly back into the pan and add the mushrooms and shallots. Sauté. Add the foie gras. Deglaze with Cognac.
- In a food processor, chop everything finely with the egg yolks, deglazed mixture, salt, pepper, and four-spice. Refrigerate the stuffing until you are ready to use it.

Simple Stuffing

6 to 8 servings

- Mix the bread and cream together.
- In a meat grinder, grind the pork, veal, chicken livers, and the soaked bread. Mix everything with the egg, salt, and pepper. Let rest for 1 hour.

Braised Walleye Fillets with Sea Beans and Strawberry Sorrel Jus

4 servings • Cooking time: 12 minutes • 2

Ask your fish monger for thick walleye fillets that won't dry out during cooking. Sea beans are a small, green, and tender annual plant. They turn red near the end of summer. They grow on seashores and in saltwater marshes.

Preparation: 30 Minutes

- Wash and blanch the sea beans twice, depending on the amount of salt. Heat the oil and sauté the onions. Add the sea beans, pepper, and cook for about 3 minutes. Set aside.
- With 3 1/2 oz (100 g) of butter, grease a baking dish where the fillets can be placed close together side by side. Place the sea beans on the bottom, then the walleye fillets. Lightly season with salt and pepper. Cover with parchment paper and bake slowly in the oven at 185 °F (85 °C). The fish is cooked when the internal temperature is 158 °F (70 °C).
- While the fish is baking, put the strawberries, sorrel, and the fish fumet into a blender. Season with salt and pepper. Blend well and adjust the seasoning. If a bitter flavour dominates, add a few more strawberries. If it is too sweet, add a few sorrel leaves.
- Remove the fish from the oven and sauce. Serve hot.

INGREDIENTS

- 7 oz (200 g) sea beans
- 1/3 cup (80 ml) sunflower oil or another neutral-tasting oil
- 2 Spanish onions, finely chopped
- Freshly ground pepper
- 5 oz (150 g) unsalted butter
- Salt and freshly ground pepper
- 4 thick walleye fillets, 5 oz (150 g) each
- 7 oz (200 g) very ripe strawberries
- 1 oz (30 g) sorrel leaves
- 6 tablespoons (90 ml) fish fumet (reduction)
- 4 oz (120 g) tomatoes, diced

INGREDIENTS

- 1 lb (480 g) ripe strawberries
- 2 cups (480 g) granulated white sugar
- 2 cups (500 ml) water
- 2 cups (500 ml) alcohol, 90%
- 2 vanilla beans
- 20 coriander seeds

Strawberry Liqueur

Preparation: 20 Minutes

- Wash the strawberries well. Drain on paper towels and hull. Crush the strawberries in a food mill. Leave for 1 to 2 hours.
- Put the granulated sugar in a saucepan and add the water. Bring to a boil and pour it immediately over the strawberry pulp. Put the mixture into a wide-mouth Mason jar. Add the alcohol, vanilla beans, and coriander seeds. Cover and keep at room temperature for 3 to 4 days.
- Strain, pour into an opaque bottle, and keep for about 2 months before consuming.

- 1 cup (250 ml) heavy cream (35%)
- 1/3 cup (80 g) sugar
- 10 Thai basil leaves, roughly chopped
- 3/4 cup (175 ml) coconut milk

Granita

- 1 lb (480 g) strawberries, washed, hulled, and cut in half
- 2 oz (60 g) sugar
- 1/2 cup (125 ml) water
- Strawberries, quartered

Coconut Milk and Basil Soup with Strawberry Granita

4 servings • Cooking time: 10 minutes • 3

Preparation: 20 Minutes

- Bring the cream and the sugar to a boil. Turn off the heat, add the basil, and infuse for 10 minutes. Add the coconut milk and gently boil again. Strain through a cone-shaped or fine mesh sieve. Chill over ice.

Granita

- In a mixing bowl, combine the strawberries and the sugar. Cover with plastic wrap. Cook for about 2 hours in a bain-marie at a gentle boil. Strain through a cone-shaped or fine mesh sieve and collect the juice. Add the water to the strawberry juice. Cool completely. Pour into a flat container and keep in the freezer.
- Pour the chilled soup into bowls. Garnish with the quartered strawberries and add the granita, which should be taken out of the freezer 2 minutes before serving.

Basic Recipes

LIGHT POULTRY STOCK

You can make this recipe with different kinds of poultry. The principle base is always the same: if you use a hen or a rooster, boil the bird whole. The longer you cook it, the more you can identify the individual flavours. If you use chicken bones, make sure to soak them to remove the impurities (blood).

- 4 1/2 lb (2 kg) poultry bones
- 10 oz (300 g) carrots, cut in medium pieces for mirepoix
- 7 oz (200 g) onions, cut in medium pieces for mirepoix
- 3 1/2 oz (100 g) leek, white part only, cut in medium pieces for mirepoix
- 3 1/2 oz (100 g) celery, cut in medium pieces for mirepoix
- 3 cloves of garlic, chopped
- 1 clove
- Black pepper
- Bouquet garni: 20 parsley stems + 1 sprig of thyme + 1/2 bay leaf

- Soak the poultry bones and drain.
- Put the vegetables, garlic, clove, and pepper into the stock pot with the soaked bones. Cover with water and bring to a boil. Skim if necessary. Add the bouquet garni. Simmer for 45 minutes. Strain through a lined cone-shaped or fine mesh sieve. Reduce if the taste isn't strong enough.

BROWN POULTRY STOCK

Some recipes require brown poultry stock. The ingredients are more or less the same as for the light poultry stock, but the method is slightly different.

- 4 1/2 lb (2 kg) poultry bones
- 2/3 cup (160 ml) cooking oil
- 10 oz (300 g) carrots, cut in medium pieces for mirepoix
- 7 oz (200 g) onions, cut in medium pieces for mirepoix
- 3 1/2 oz (100 g) leek, white part only, cut in medium pieces for mirepoix
- 3 1/2 oz (100 g) celery, cut in medium pieces for mirepoix
- 3 cloves of garlic, chopped
- 1 clove
- Black pepper
- 1 bouquet garni
- Tomato paste (optional)

- Chop up the bones with a cleaver and brown in the oven on a cookie sheet with 5 tablespoons (75 ml) of oil, until golden (this will give colour to the stock).
- Sweat the vegetables in the rest of the oil. Put them in the stock pot with the bones and seasonings. Cover with water and simmer for 45 to 60 minutes. If the stock isn't brown enough, add a bit of tomato paste. Strain through a lined cone-shaped or fine mesh sieve.

BROWN VEAL STOCK

Never add salt to a stock. Since it needs to be reduced, it will end up being too salty. Skim frequently while the stock simmers. If it evaporates too much, add water. After it is finished cooking, strain the stock through a lined cone-shaped or fine mesh sieve. Let it cool then fill small containers that can be frozen. Stocks can be made in winter for later use. They smell wonderful and add humidity to the air in the house. To quicken the cooling process, pour the stock into a bowl, put it in the sink and slowly run cold water around it. If the stock is fatty, skim it frequently while it cooks.

- 22 lb* (10 kg) veal bones (preferably knees, chopped into small pieces by your butcher)
- Vegetable shortening
- Vegetable oil
- 2 lb (1 kg) onions, cut in large pieces for mirepoix
- 2 lb (1 kg) carrots, cut in large pieces for mirepoix
- 1 lb (500 g) celery, cut in 2" (5 cm) pieces for mirepoix
- 2 heads of garlic, unpeeled
- 1 bay leaf
- 2 sprigs of fresh thyme
- 7 oz (200 g) parsley
- 25 black peppercorns
- 7 oz (200 g) tomato paste

- Heat the vegetable shortening in a roasting pan in the oven at 400°F (200°C). When it is nice and hot, add the veal bones and roast until they are golden on all sides. This step is very important as the roasted juices will give a nice colour to the veal stock.
- In a saucepan, sweat all the vegetables in the vegetable oil. Add the garlic, seasonings, and tomato paste. Simmer.
- When these two steps are completed, mix the two preparations together in a stock pot that is large enough to hold everything. Cover completely with water and let simmer for at least 6 hours.
- By reducing it and concentrating the juices, you obtain a demi-glace, and by reducing it even more, a veal glace. This stock isn't thickened, but by adding white roux, the stock becomes a thickened brown stock.

* This quantity is ideal, but you might not have a stock pot large enough. If so, you can divide it into smaller portions.

WHITE AND BROWN ROUX

A roux is preferred over starches, as the gluten in flour binds sauces much better.

- 3 1/2 oz (100 g) butter
- 3 1/2 oz (100 g) flour

- Melt the butter in the microwave for 20 seconds and add the flour. Continue to cook in 20-second intervals, stirring well after each interval. The roux is finished cooking once it starts to bubble.
- Note: For a brown roux, follow the same steps as with the white roux, but continue to cook it until it turns brown.

ROUX SUBSTITUTES

There are store-bought substitutes available to bind or thicken stocks and sauces. The most commonly used is corn starch. If corn starch is used to bind a sauce, it should be served immediately. If not, it will thin out or separate within about 20 minutes. The results are the same with all the starches (potato, rice, arrowroot, chestnut, etc.) The advantage of thickening with rice or potato starch is that they don't leave a residual taste. There are also a variety of store-bought thickeners such as Veloutine, among others.

BLUEBERRY OR CRANBERRY SAUCE

If the sauce is not thick enough, add white roux or a corn starch mixture as thickener. At the last minute, you can also add a few drops of alcohol made from the same fruit.

- 7 oz (200 g) unsalted butter
- 4 dried shallots, chopped
- 7 oz (200 ml) tannic red wine
- 1/2 cup (125 ml) blueberry juice
- 1 cup (250 ml) brown game-meat stock
- 1 cup (120 g) blueberries, fresh or frozen
- Salt and pepper

- Heat half of the butter and sweat the shallots. Add the wine and blueberry juice. Reduce by 90%. Add the game-meat stock, simmer for 10 minutes, and strain through a cone-shaped or fine mesh sieve. Set aside.
- While the sauce simmers, sauté the blueberries in the rest of the butter until they burst. Place them on paper towels. A few minutes before serving, add the blueberries to the sauce and adjust the seasoning.
- The same recipe can be made by replacing the blueberries with cloudberries.

RED CURRANT SAUCE

Red currants are small, red, soft, and very tart. In Canada, they are grown in the Niagara region. They can be readily replaced with wild cherries.

- 2 dried shallots, finely chopped
- 2 cups (500 ml) red wine
- Cherry juice
- A pinch of ground cinnamon
- 7 oz (200 ml) thickened brown game-meat stock
- 6 tablespoons (90 g) butter
- 3 1/2 oz (100 g) fresh breadcrumbs made from sliced white bread
- 7 oz (200 g) Morello cherries, pitted and canned
- 2 tablespoons (10 g) lemon zest, chopped
- Salt and pepper

- Put the shallots, wine, cherry juice, and cinnamon into a saucepan. Reduce by 90%. Add the game-meat stock. Adjust the seasoning and strain through a lined cone-shaped or fine mesh sieve.
- Just before serving, finish the sauce with butter. Add the breadcrumbs, Morello cherries, and lemon zest.

GAME-MEAT SAUCE

- 1/3 cup (60 g) carrots, diced
- 1/3 cup (60 g) celery, diced
- 1/3 cup (60 g) shallots, chopped
- 1/4 cup (60 g) unsalted butter
- 1 1/2 lb (600 g) game-meat trimmings
- 2/3 cup (160 ml) white wine
- 1 1/4 cups (300 ml) thickened game-meat stock, reduced
- 1 bouquet garni
- 3 cloves of garlic
- 1/4 cup (60 ml) duck or other animal blood
- 2 1/2 tablespoons (37 ml) Armagnac or Cognac
- Salt and pepper

- Sweat the carrots, celery, and shallots in the unsalted butter. Add the game-meat trimmings until they become firm. Add the wine and reduce by 90%.
- Add the game-meat stock, bouquet garni, and garlic. Simmer slowly for 30 minutes.

- Strain through a cone-shaped or fine mesh sieve and adjust the seasoning. Finish by binding with the blood and Armagnac.

ORANGE SAUCE

- 2/3 cup (160 ml) orange juice, freshly squeezed
- 6 tablespoons (90 ml) white wine
- 1/4 cup (60 ml) cider vinegar
- 1/2 cup (125 g) granulated white sugar
- 1 1/3 cups (320 ml) thickened brown game-bird stock
- 1/3 cup (80 ml) Grand Marnier
- Salt and pepper

- Mix the orange juice, wine, and the sugar. Cook until caramelized. Stop the cooking by pouring a little cold water into the caramel (be careful of splashing—it's very hot!).
- Heat the game-bird stock and add it to the orange juice mixture. Adjust the seasoning and add the Grand Marnier. You can also add blanched orange zest.

SAUCE *GRAND VENEUR* (GAME-MEAT)

Real grand veneur sauce (literally, huntsman's sauce) is made with game-meat blood. This is a variation. If you shoot the game with bullets, the muscles will retain blood, which will affect the sauce. The famous chef Auguste Escoffier added cream to this sauce.

- 14 oz (400 ml) game-meat marinade with vegetables
- 10 oz (300 ml) thickened brown game-meat stock
- 3 1/2 oz (100 g) butter
- 2 tablespoons (30 ml) red currant jelly
- 6 tablespoons (90 ml) Cognac or Armagnac
- Salt and pepper

- Reduce the marinade with vegetables by 90%. Add the game-meat stock. Cook for 10 minutes and strain through a cone-shaped or fine mesh sieve. Finish with butter, red currant jelly, and Cognac or Armagnac. Adjust the seasoning.

VENISON SAUCE

- 3 1/2 oz (100 g) unsalted butter
- 2 dried shallots, chopped
- 13 oz (400 g) game-meat trimmings
- 3 cups (750 ml) red wine
- 3 cups (750 ml) sauce *poivrade* (a peppery brown sauce made with wine, vinegar, and cooked vegetables, which is strained before serving)
- Salt and pepper

- Heat the butter. Add the shallots and sauté slowly. Add the meat trimmings, sear, and add the wine. Reduce by 90%. Add the sauce *poivrade*. Cook for 20 to 30 minutes and strain through a cone-shaped or fine mesh sieve. Adjust the seasoning.
- You can add 1 tablespoon of red currant jelly and 2 1/2 tablespoons of port wine.

SWEET AND SOUR SQUASHBERRY SAUCE

- 1 cup (160 g) squashberries, fresh or frozen
- 2/3 cup (160 g) granulated white sugar
- 1 1/4 cups (300 ml) thickened brown game-meat stock
- 1/2 cup (80 g) celery, diced
- 2 oz (60 g) squashberries
- 3 1/2 oz (100 g) unsalted butter
- Salt and pepper

- Juice the squashberries. Add sugar and caramelize. Stop the cooking by pouring in a little cold water, but be careful not to burn yourself when it splashes. Heat the game-meat stock. Add the diced celery and cook for 15 minutes on low heat. Add the reduced juice, squashberries, and finish with butter. Adjust the seasoning.

FISH FUMET

This fumet (reduction) keeps in the freezer for a maximum of 2 to 3 months. Avoid using carrots when making fish fumet, as they generally give a sweet taste to the stock. Never salt a fish fumet because you might want to reduce it at some point to get a fish concentrate.

- 1 1/2 tablespoons butter
- 1 3/4 lbs (800 g) fish bones and trimmings (preferably from flatfish)
- 2 1/2 oz (75 g) onions, minced
- 4 oz (125 ml) leeks, minced
- 4 oz (125 ml) celery, minced
- 6 tablespoons (90 g) shallots
- 5 oz (150 ml) mushrooms, minced
- 6 tablespoons (90 ml) white wine
- 4 teaspoons (60 ml) lemon juice
- 4 cups (1 litre) cold water
- A pinch of thyme
- 1/2 bay leaf
- 10 peppercorns

- Heat the butter in a saucepan, add the fish bones and trimmings and all the vegetables. Sweat for 4 to 5 minutes. Moisten with wine, lemon juice, and cold water. Add the thyme, bay leaf, and pepper. Bring to a boil. Simmer for 25 minutes. Strain through a cheesecloth, cool, and keep until needed.

HOLLANDAISE SAUCE

Always use unsalted butter because of the higher fat content.

- 6 tablespoons (90 ml) white wine
- 5 tablespoons (75 g) shallots, chopped
- 2 teaspoons (10 ml) white vinegar
- 4 egg yolks
- 1 cup (240 g) unsalted butter
- Salt and pepper
- Juice of half a lemon (optional)

- Reduce the wine and vinegar with the shallots by 90%. Cool this reduction and add the egg yolks. Strain through a lined cone-shaped or a fine mesh sieve.
- In a round bowl (a double boiler for restaurateurs) that can be heated, whisk the mixture, and add the salt and pepper. Over warm water or in a bain-marie, beat the mixture until it forms ribbons (as with whipped cream). This step is very important. Beating the yolks with the acidity of the white vinegar ensures the success of this sauce. At the same time, melt the butter. When the sauce forms ribbons, beat in the melted butter a little at a time. The mixture should be smooth. Add the lemon juice if necessary.

LOBSTER SAUCE

- 2 lbs (900 g) lobsters, or lobster shells
- 3 tablespoons (45 ml) olive oil
- 1/4 cup (60 g) unsalted butter
- 2 tablespoons (30 g) shallots, chopped

- 1/4 teaspoon (1 g) garlic, chopped (centre sprout removed)
- 1/2 cup (125 ml) cognac
- 6 tablespoons (90ml) white wine
- 3 1/2 cups (850 ml) fish fumet
- 2 tablespoons (30 ml) tomato paste
- 1 tablespoon (15 g) parsley, roughly chopped
- 1/2 teaspoon (2.5 g) Cayenne pepper
- 1/2 teaspoon (2.5 g) salt

- Cut the lobster tails into pieces and break the claws. Split the body in half lengthwise. Remove the sand pouch, situated close to the head. Set aside the creamy parts and the meat.
- In a sauté pan, heat the oil and butter. Get a good sear on the pieces of shell to get a nice red colour. Remove the excess fat and add the other ingredients. Cook covered at 400°F (200°C) for about 20 minutes. Drain the pieces of shell and crush them into the sauce along with the creamy parts and the lobster meat. Cook on high heat to reduce while whisking. Strain through a cone-shaped sieve lined with cheese cloth or a fine mesh sieve. Set aside in the fridge until you're ready to use it.

FISH MOUSSE

- 2 lbs (1 kg) pike, flounder, or Pollock (cleaned)
- 4 or 5 egg whites
- Salt and ground pepper
- Nutmeg
- 4 cups (125 ml) heavy cream (35%)
- 2 cups (480 g) unsalted butter

- Crush the fish with a mortar and pestle. Add the egg whites and seasonings. Pass through a fine sieve and set in a sauté pan on ice to rest for about 2 hours.
- Working carefully with a wooden spoon, progressively dilute the mixture with cream and butter while over ice. Let rest overnight in the fridge.

MAYONNAISE

- 4 egg yolks
- 1 tablespoon (15 ml) Dijon mustard
- Salt and white pepper
- High quality white vinegar
- 4 cups (1 litre) oil of your choice (peanut, canola, corn, hazelnut, walnut, olive, sunflower, pistachio, etc.)

- In a blender or with a whisk, mix the egg yolks with the mustard, salt, pepper, and a few drops of vinegar. It is important to put the salt in at this point so it dissolves. Drizzle in the oil a little at a time. If the mixture becomes too firm, add a few drops of vinegar or water, then continue adding the oil.

GRIBICHE SAUCE

Gribiche sauce is a mayonnaise made with capers, cornichons, hard-cooked eggs, and herbs.

- 4 teaspoons (20 ml) Dijon mustard
- 6 egg yolks
- 6 tablespoons (90 ml) red wine vinegar
- 3/4 cup (375 ml) olive oil
- 1 teaspoon (5 g) parsley, chopped
- 2 oz (60 g) sour pickles (cornichons), chopped
- 1 teaspoon (5 g) chervil, minced
- 1/2 teaspoon (2.5 g) tarragon, chopped

- 2 oz (60 g) capers, chopped
- 3 egg whites, hardboiled and chopped
- Salt and pepper

- Beat the mustard with the egg yolks. Add the vinegar and the olive oil. Add the parsley, pickles, chervil, tarragon, capers, and the hardboiled egg whites. Season to taste with salt and pepper.

PÉRIGUEUX SAUCE

Périgueux sauce is recommended for filet of beef, veal, fois gras or eggs.

- 1 1/2 oz (45 g) shallots, very finely chopped
- 1/2 cup (120 g) unsalted butter
- 2/3 cup (160 ml) red port
- 3/4 cup (175 ml) Madeira
- 4 cups (1 litre) thickened game-bird or -meat stock
- 1/4 cup (60 ml) truffle juice
- Black truffle, thinly julienned

- Sweat the shallots gently in half of the butter. Add the port and Madeira. Reduce by 90%. Add the game stock. Simmer for 10 minutes and strain through a lined cone-shaped sieve. About 15 minutes before serving, finish the sauce with the rest of the butter. Add the truffle juice and julienned truffles.

UNCOOKED MARINADE

The liquid should never top more than a third of the meat. Cover with plastic wrap or a lid and keep it in a place that is neither too hot nor too cold. Flip the piece of meat over twice a day.

- 4 cups (1 litre) tannic red wine
- 2/3 cup (160 ml) red wine vinegar
- 2/3 cup (160 ml) grape seed oil
- 3 shallots, finely minced
- 1 onion, finely minced
- 6 juniper berries
- 12 black peppercorns
- 8 carrots, in very thin rounds
- 1 celery stalk, finely minced
- 2 cloves of garlic
- 2 bay leaves
- 10 parsley sprigs
- 1 sprig of fresh thyme
- Salt and pepper

- Mix all the ingredients. Leave for 5 to 6 hours on the counter. Pour it over the meat to marinate.

Why Use a Thermometer for Cooking?

When I began in this profession forty-eight years ago, the chef taught me to tell the temperature of cooked meat or fish "by touch." This method could never be precise since, depending on the quality of the meat and its age, there could be a big difference in how well it's cooked. Today, the thermometer is indispensable in controlling the temperature of meat or fish, or in knowing the real temperature of your oven. That's why I rarely specify cooking times.

To better understand the usefulness of a thermometer, it's enough to remember that the fear of salmonella that tormented our parents and grandparents caused them to overcook chicken to kill harmful bacteria. It was the same for pork and beef because cooking destroyed tapeworm and its eggs. In a way, they were right to overcook it. Henhouse, pigsty, and stable hygiene were not a priority for farmers during the last century, but today we know how to control parasites and bacteria.

Dr. Pierre Gélinas, in his index of pathogenic microorganisms transmitted by food (*Répertoire des microorganismes pathogènes transmis par les aliments*), shows us how to ensure that the culinary arts are both celebratory and safe. It is no longer necessary to overcook roast pork, chicken, or a piece of beef.

Because of this, food is much more delicious.

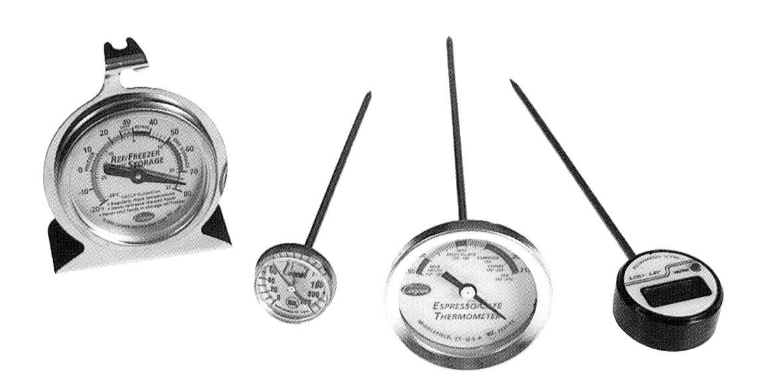

Correct Cooking Temperatures

Food	Cooking Temperature	Internal Temperature	Effects
Beef	Maximum Temperature: 158 to 248 °F (120 to 70°C)	122 °F (50 °C) 125 °F (52 °C)	Shrinks the flesh Kills parasites and bacteria
Veal	Minimum Temperature: 129 to 154 °F (54 to 68 °C)	129 °F (54 °C)	Rare for red meats
Pork	Depending on the cooking method, the temperature will be higher or lower	131 °F (55 °C)	Proteins break down
		133 °F (56 °C)	Pink centre in red or white meats
		136 °F (58 °C)	Done “à pointe”
		144 °F (62 °C)	From rare to cooked (alters the chemical structure of the albumin)
		151 °F (66 °C)	Irreversible colour, alters the chemical structure of the myoglobin, coagulation of collagen
		154 °F (68 °C)	Loss of water retention
		174 °F (79 °C)	Essential coagulation of proteins
Chicken white meat	162 °F (72 °C)	151 °F (66 °C)	
Lamb	172 °F (78 °C)	118 °F (48 °C)	
Fish	176 °F (80 °C)	126 °F (52 °C)	

Be Aware of Bacteria and Parasites

<table>
<tr><th>Pathogenic microorganism or parasite</th><th>Foods they are found in</th><th>Cooking temperature to render harmless</th></tr>
<tr><td>Salmonella</td><td>Eggs, unpasteurized milk products, raw milk cheeses, raw fish products (particularly oysters, mussels, scallops), meats (especially poultry, but also pork), fruits and vegetables washed in contaminated water</td><td>122 °F (50 °C)
125 °F (52 °C)</td></tr>
<tr><td colspan="3">Salmonella cannot survive cooking at 150 °F (66 °C) lasting 12 minutes.</td></tr>
<tr><td>TAPEWORM
Taenia saginata (beef)
Taenia solium (pork)</td><td>Raw or insufficiently cooked pork or beef
Tapeworm is a parasite found in these animals</td><td>Pasteurization and freezing destroy the larva of these parasites</td></tr>
<tr><td colspan="3">It is better to freeze beef before making steak tartare.</td></tr>
<tr><td colspan="3">Keep food in the refrigerator, between 0 and 4 °C.
Never leave perishable food for a long time in a car or on the kitchen counter as salmonella develops between 5 and 12 °C.
Be wary of cooking things in the microwave, and also of some salted or smoked foods.
To avoid botulism, never keep garlic in oil and immediately consume preserves made with mushrooms, corn, and beans.
If you take all the necessary precautions to cook and preserve food, you will discover a different way of cooking.</td></tr>
</table>

Glossary

"À GOUTTE DE SANG"

When the breast of poultry or game bird is just done, pierce it with a skewer. A drop of blood should pearl in the centre of the fat.

AIGUILLETTE

Long, thin slices of duck breast.

AROMATICS

All herbs and plants or roots that release pleasant smells.

BARD OR BACK FAT

Streaky bacon.

(TO) BARD

To wrap streaky bacon or fat around poultry, game meat, or a roast to keep it moist during cooking.

BLANCH

Bring food to a boil and then shock it in cold water to maintain the colour. It can also mean to remove the skin from tomatoes, almonds, hazelnuts, etc., by immersing them in boiling water for 10 to 15 seconds, then shocking them in cold water.

BOUQUET GARNI

A bundle of thyme, bay, celery, and parsley sprigs tied together. It adds a pleasant aroma to dishes.

BRAISE

Cook in a small amount of liquid in a braising pan or a pot with an airtight lid. Since this method of cooking takes a long time, use good, airtight equipment to prevent evaporation of the cooking liquid.

BREAD

Soak food in beaten egg yolks, then in fresh or dried breadcrumbs.

BREADCRUMBS

Stale bread crusts pushed through a flat sieve and dried. Fresh bread without crusts can also be used.

BROWN

Sauté quickly in hot oil meats, game meats, or a vegetable that you want to give an even colour to before adding liquid.

BROWNING BONES

To roast or brown in fat the bones of game meat and fowl in the oven.

CAUL FAT

A fine, fatty membrane that covers the intestines of veal and pork. It is used to hold together a ground or minced preparation while it cooks.

COAT

To cover hot and cold dishes with a sauce or a jelly.

COOKING OIL

Peanut, canola, corn, olive, sunflower.

CRUSH or GRIND

Roughly chop, usually tomatoes, or use a mortar and pestle.

CUTLET

Cut pieces of meat on the bias into thin slices. A wooden mallet can be used to pound the raw cutlets (escalopes) into uniform pieces.

DECANT

To separate the liquid from the other ingredients cooked in a sauce.

DICE

Vegetables cut into small cubes of 1 to 3 mm.

DUST WITH FLOUR

Sprinkle with flour at the beginning of cooking to bind with a liquid.

FINE DICE

Cut thin strips of vegetables like chives, lettuce, sorrel, etc., into tiny pieces.

FINISH A SAUCE WITH BUTTER

When a sauce or a jus is done, take it off the heat and stir in knobs of butter (72 °F / 22 °C) without bringing it to a boil. The butter binds the sauce together.

GASTRIQUE

Sweet and sour mixture that serves as a base for some sauces.

INSERT (CONTISER)

Cut incisions into poultry, game meats, or some fish into which slices of truffles or other ingredients are stuffed.

INTERNAL TEMPERATURE

Degree of doneness at the centre of a piece of meat.

JULIENNE

Cut into thin strips of 3 to 5 cm in length and 1 to 2 mm thick.

LARD

Using a larding needle, insert strips of pork fat into meat to keep it moist. It is used mostly with drier cuts of meat (for example, the outer parts of the thigh).

LARDING NEEDLE

A small instrument that resembles a large needle. It is used to thread strips of fat into a piece of meat.

MACERATE

Soak fruit, vegetables, or meat in alcohol or an aromatic liquid.

MARINATE

Soak in a marinade to tenderize food and give it more flavour.

MELT

See Render.

MINCE

Cut food into thin slices, especially vegetables like onions, leeks, etc.

MIREPOIX

A mixture of diced carrots, onions, celery, and sometimes lard or ham that serves as a base for sauces.

MIXTURE

A combination of one or more ingredients, to be used on its own or added to another mixture.

MOISTEN

Add water, stock, or consommé to various ingredients to help cook them.

MOUNTAIN OYSTER

Animal testicles.

PIT
Remove the seeds from fruit or vegetables.

PLUCK
Pick the leaves in sprigs from herbs such as chervil and parsley.

PREPARATION
Step by step instructions for cooking a dish.

REDUCE
Boil or simmer a sauce or stock to make the flavour stronger.

RENDER
Cook to soften in butter and water until the liquid has completely evaporated.

SALPICON
Various ingredients diced, in small slices, or fillets bound together with a sauce. Often used in hors d'oeuvres or stuffings.

SAUCEPAN
A pan with sloping sides usually used to make sauces.

SAUTÉ
Cook meat quickly in fat in a frying pan or sauté pan.

SAUTÉ PAN
A flat-bottom pan with straight sides.

SEAL
A paste made of flour mixed with an egg white or water used to seal the rim of a lid to a pot.

SEAR
To cook meat quickly on both sides in very hot fat in a frying pan on high heat.

SET ASIDE
Keep an ingredient or a dish before using.

SIEZE
Seal the surface of meat without browning it.

SIMMER
Cook gently and slowly.

SKIM
Cook for a long time, occasionally removing the fat from the surface.

SMOOTH
Prevent the formation of a skin on a sauce or custard by gently stirring it or mixing in softened butter.

SOAK
Soak meat or vegetables in cold water to remove the impurities.

STEW
To cook food in a sealed Dutch oven, a vacuum bag, or in a clay pot.

STUFF
Fill the inside of poultry or a piece of meat with a stuffing.

SWEAT
To cook vegetables in fat until they release some of their liquid.

Index